Best wishes!

Juval

STAYING
SAFE

HarperResource

An Imprint of HarperCollins*Publishers*

- Protect Against Fraud

- Travel Safely Abroad

- Safeguard Your Identity

Identify Potential Threats • Become Security Minded

THE COMPLETE GUIDE TO PROTECTING YOURSELF, YOUR FAMILY, AND YOUR BUSINESS

STAYING SAFE

JUVAL AVIV

**Former Israeli
Counterterrorism
Intelligence Officer**

A hardcover edition titled *The Complete Terrorism Survival Guide* was published in 2003 by Juris Publishing, Inc.

HarperCollins books may be purchased for educational, business, or sales promotional use. For information please write: Special Markets Department, HarperCollins Publishers Inc., 10 East 53rd Street, New York, NY 10022.

Library of Congress Cataloging-in-Publication Data

Aviv, Juval.
    [Complete terrorism survival guide]
Staying safe : the complete guide to protecting yourself, your family, and your business / Juval Aviv.
        p. cm.
    "A hardcover edition titled The complete terrorism survival guide was published in 2003 by Juris Publishing, Inc."—Verso t.p.
    Includes bibliographical references and index.
    ISBN 0-06-073520-1
    1. Terrorism—Prevention. 2. Crime prevention. 3. Dwellings—Security measures. 4. Travel—Safety measures. 5. Survival skills. I. Title: Complete guide to protecting yourself, your family, and your business. II. Title.
HV6431.A95 2004
613.6'9—dc22   2004047590

07 08 DIX/RRD 10 9 8 7 6 5

To Tsila, Atalya, Don, Todd, and Zoé
for their immeasurable love and support.

ACKNOWLEDGMENTS

This book has been in the making for many years. Yet, the events of September 11 underscored the urgent need for it to be written and published. A number of people have shown considerable support for this project. This book would not have been written without their sustained interest and guidance.

I would like to thank Anthony Williams, Esq., of Coudert Brothers for motivating me to put my thoughts down on paper.

I am deeply grateful to Daniel Aharoni, Esq., my friend and general counsel, for continually getting me out of trouble.

I am indebted to Lawrence Newman, Esq., of Baker and McKenzie for his steadfast support and meaningful insights.

And finally, to my staff at Interfor, Inc., for their tireless efforts. A special thanks to Anna Moody, Esq., and Abby Barasch for taking charge of this project.

CONTENTS

PART THREE ✚ SAFEGUARDING YOUR IDENTITY AND PROTECTING YOUR BUSINESS

INTRODUCTION

Let's get something straight from the get-go: a person *chooses* to stay safe.
Many victims of crime and even terrorism became victims because they made
poor choices.

"How could that be?" you may be asking. "What poor choices did the victims of 9/11 make?"

Let's consider just the occupants of New York's World Trade Center. In
1993, a terrorist bomb exploded in the complex's underground garage, killing
six and wounding a thousand. The perpetrators had hoped to topple the
south tower. They failed at that, of course. But the 1993 attack put all World
Trade Center occupants on notice that they worked at an address targeted by
terrorists. Armed with that knowledge, certainly some tenants moved out of
the towers after the bombing, some workers changed jobs to avoid a repeat,
and some prospective workers simply never applied for or accepted jobs located in the towers.

This is what we mean when we say that staying safe is often a matter of
personal choice.

The information contained in this book will help you in making the right
choices. The material is logical, straightforward, and easy to follow. You don't
need to have the strength of a Navy SEAL or the dexterity of a James Bond to
take advantage of our advice. And you should learn how to stay safe not just
because of the events of 9/11 but also because you never know what dangers
you may confront in this crazy world.

Our method of conveying this information to you—and making it part of
your daily life—is simple and effective. You might call it osmosis. In science,
this occurs when a solvent passes from one side of a membrane to the other. It
takes place all the time in the human body, as substances pass through cell

walls and are absorbed. In the realms of the mind and emotions, osmosis occurs when we absorb new ideas, feelings, or attitudes, seemingly without effort. These new concepts needn't be drilled into us. Instead, they become part of us, changing us, by our mere exposure to them.

SELF-RELIANCE NOW MORE THAN EVER

Crime, of course, is as old as humanity itself. It's an expression of human nature, arising from greed, avarice, hatred, jealousy, and other base emotions. In response, we've developed defenses that go by the general name "common sense." We continually make decisions—consciously and unconsciously—to help ensure our safety. It's a simple fact of everyday life.

But today's threats to our person and property are such that common sense can no longer be our sole protector. Staying safe now requires proactive measures.

Consider the modern crime of identity theft. An identity thief can rack up huge debts in your name, declare bankruptcy in your name, and even commit crimes using your name. Trouble is, you probably won't know about any of this until months, perhaps years, later. By that time, your credit and reputation may be ruined.

Staying safe also took on new urgency and meaning with the tragic events of September 11. America lost its innocence that day. No one will ever forget the woolly black smoke that issued from the upper floors of the World Trade Center, the hellish fireballs that erupted as the planes struck, or the surreal collapse of the two 110-story towers into gray tidal waves of dust and debris. But more was on fire that morning than just the Twin Towers and the Pentagon. Also aflame was the implicit contract Americans had with their government to keep them safe from foreign aggressors.

One lesson of September 11 is that no one should depend solely on government for his personal protection. Security is now a matter of personal initiative. To be protected against terrorism—or other forms of violence— means that individuals must become proactive, taking steps to prevent attackers from getting to them and developing a mind-set that makes security precautions a habit, as much a part of daily life as brushing your teeth. Companies, too, have to make security a higher priority, because business has been the terrorists' number one target over the years.

This marks a break with long-standing tradition. From time immemorial, people have banded together for their mutual protection. There is, after all, safety in numbers. Our primeval ancestors discovered it was easier to safeguard their lives and property when they joined to fashion a common defense against aggressors than to try to go it alone. From this native desire for self-preservation sprang the first civilized communities, cities, and governments.

Noted English political philosopher John Locke (1632–1704) explained it well when he wrote that fear of invasion and the theft of property led to the organization of societies and the formation of governments. A person living in the "state of nature" before governments were formed was free of all restraints yet "constantly exposed to the invasion of others." This meant, Locke said, "the enjoyment of the property he has in this state is very unsafe, very insecure. This makes him willing to quit this condition which, however free, is full of fears and continual dangers; and it is not without reason that he seeks out and is willing to join in society with others who are already united, or have a mind to unite for the mutual preservation of their lives, liberties and estates, which I call by the general name—property."

Nearly a century later, these words inspired Thomas Jefferson to write in the Declaration of Independence: "We hold these Truths to be self-evident, that all Men are created equal, that they are endowed by their Creator with certain unalienable Rights, that among these are Life, Liberty, and the Pursuit of Happiness—That to secure these Rights, Governments are instituted among Men."

Today's terrorists mean to break the age-old link between mutual defense and personal safety. Terrorists try to avoid clashing with a nation's military head-on; conventional warfare isn't their forte. Instead, they attack a country's cherished symbols, target its businesses, and murder its innocent civilians. Their hope is to cast doubt on a government's legitimacy by making a terrified people ask, "If our government can't protect us from harm, what good is it?"

As British prime minister Tony Blair stated so eloquently the day after September 11, "The world now knows the full evil and capability of international terrorism, which menaces the whole of the democratic world. The terrorists responsible have no sense of humanity, of mercy, or of justice. To commit acts of this nature requires a fanaticism and wickedness that is beyond our normal contemplation."

Blair's concern is well-founded. Terror groups like Osama bin Laden's Al Qaeda have expressed a keen interest in acquiring weapons of mass destruction (WMD)—that is, chemical, biological, radiological, and nuclear weapons. "What makes a WMD terrorist incident unique is that it can be a transforming event. A terrorist attack involving weapons of mass destruction would have catastrophic effects on American society beyond the deaths it might cause," Frank J. Cilluffo, codirector of the Terrorism Task Force at the Center for Strategic and International Studies in Washington, D.C., advised a House panel in October 1998. "Aside from the actual physical effects and human suffering resulting from a WMD event, the psychological impact would be enormous, shaking the nation's trust and confidence in its government to its core."

Our safety, in essence, is determined by the barriers we have in place. These include not only physical barriers (e.g., doors and locks) but also encompass behavioral barriers (e.g., personal habits and routines) and systems barriers (e.g., security personnel and computer firewalls). You might think of these barriers in terms of concentric rings, with you, your loved one, and your business at the center. The outer rings aim to protect you against the more distant threats, while the inner ones are intended to repel more immediate dangers.

Staying safe thus means designing a combination of systems, procedures, and physical roadblocks that minimize risk and exposure to both crime and terrorism.

Security is a function of logic, and a belief in its ability to keep us safe is an essential prerequisite. But believing in security is only half the battle, for staying safe also entails keeping an open mind and planning ahead. Here, knowledge is vital. Travelers, for example, have to learn more about their destinations prior to departure and plan accordingly.

The primary goal of safety and security planning must be prevention. As the saying goes, an ounce of prevention is worth a pound of cure. The more numerous and sophisticated the barriers erected, the harder and longer it will be for an intruder to get past them. A criminal or terrorist may give up after deciding his intended target is too hard to attack and go on to someone or someplace else. The philosophy of security also means planning for the unexpected. Israel, for instance, has had contingency plans in place for decades to deal with hijacked-aircraft attacks on tall buildings and city centers. Its security forces are prepared for such an emergency.

Staying safe, furthermore, is as much an attitude as it is investment. For corporate security to be effective, for instance, management from top to bottom must subscribe to a philosophy of security. A view toward prevention must infuse a company. Additionally, it isn't enough merely to protect a few senior executives from assault, kidnapping, assassination, and the like. Rank-and-file personnel must also be protected. And that doesn't mean giving them keys to newly locked washrooms. Frankly, if management doesn't subscribe to the whole package and doesn't embrace a proactive philosophy of security, it oughtn't to take any precautions at all. There's no such thing as being half-pregnant. Businesses can't cherry-pick security measures and expect to be safe.

THE PANOPLY OF HAZARDS

The potential threats we face today are many. They run the gamut from the ordinary (e.g., fraud, burglary, and assault) to the extraordinary (e.g., cyberattacks, identity theft, and terrorism). Indeed, what we now know of international terrorism can lead to only one conclusion: if we rule out the possibility of more barbaric attacks in the future, we do so at our peril.

Consider the nerve-gas assault on the Tokyo subway system in March 1995. Carried out by members of the Japanese doomsday cult Aum Shinrikyo, the gas killed 12 persons and injured 5,700 others. Until that happened, few thought the deployment of nerve gas by terrorists was really possible. "The Aum Shinrikyo incident provides a poignant example of how old notions of threats can restrict our scope of vision, causing us to miss important new threats," observed John V. Parachini, a senior associate at the Monterey Institute of International Studies Center in California.

Here's a brief rundown of some of the potential hazards we face today.

CHEMICAL WEAPONS Chemical munitions are among the most feared weapons in the world, especially because the body's reaction to them can be so acute, with death often coming in minutes. Toxic chemical weapons fall into four categories: nerve agents, blister or vesicant agents, blood agents, and pulmonary agents. Nerve agents, such as sarin, soman, tabun, and VX gas, can incapacitate a victim in an instant and kill within 15 minutes. Blister agents, such as mustard gas, are slower acting and less deadly. Blood agents, such as hydrogen cyanide, rapidly cause seizures, respiratory failure, and cardiac arrest. And pulmonary agents, such as chlorine and phosgene, vary in effect but are often lethal. Heavy metals such as arsenic, lead, and mercury also have weapons potential.

The release of gas canisters is the most likely method of attack, but it's not the only way of dispensing poison gas. One weapons expert has postulated that terrorists dressed as a maintenance crew might enter a train station, subway, large building, or airport with drums labeled as cleaning fluid. The drums would, in fact, contain a deadly poison, which could, if unleashed at rush hour, cause tremendous casualties.

An indirect means of attack would be to blow up a chemical factory or chemical-filled railcar. Deadly chemicals produced commercially in the United States and most industrial economies include chlorine and hydrogen fluoride, among others. Even phosgene, the deadly gas first used in combat in World War I, can be found at more than 30 U.S. chemical plants. All told, approximately 850,000 facilities in the United States use hazardous chemicals.

BIOLOGICAL WEAPONS Biowarfare has a long pedigree. One of the first instances took place in the sixth century BC when Assyrians tainted an enemy's wells with rye ergot, a convulsion-causing fungus. In 1346, the Tartar Mongols started an epidemic in the besieged Crimean town of Kaffa by hurling the corpses of plague victims over the city's walls. In North America, during the French and Indian Wars of 1754–63, British troops were suspected of distributing smallpox-infected blankets to Native Americans. And the Japanese used biological agents in World War II against the Chinese and in grisly experiments on prisoners of war. More recently, an Ohio microbiologist, said to have suspect motives, was arrested in 1995 after fraudulently acquiring by mail the bacterium responsible for pneumonic plague (i.e., the black death). Also in 1995, Japan's Aum Shinrikyo cult, besides disbursing sarin gas in the subway, made at least nine attempts—all of which were unsuccessful—to infect central Tokyo with aerosolized anthrax and botulism. The cult further tried to obtain the deadly Ebola virus.

In fact, of all the weapons of mass destruction, biological agents pose the worst threat. Our assessment is based on the ease with which a determined terrorist organization could acquire biological-weapons technology, materials, and expertise and the impact a large-scale attack would have on a nation's population and economy. A nuclear blast might kill and injure more people in a single incident, but bioterrorism presents the frightening prospect of a virtually nationwide—or even a multinationwide—barrage that could inflict innumerable casualties. As early as 1984, in fact, terrorism experts Neil C. Livingstone and Joseph D. Douglass Jr. warned that chemical-biological weapons (CBW) were "the poor man's atomic bomb."

Biological agents that could be used against civilian populations include anthrax, botulism, hemorrhagic fever viruses (including the deadly Ebola virus), plague, smallpox, and tularemia. (For detailed information on these bioagents, see the Centers for Disease Control and Prevention Web site at www.cdc.gov.) Another nasty toxin is ricin. Derived from the beans of the castor plant, it's one of the most lethal naturally occurring toxins known. Castor beans are available worldwide, and ricin is fairly easy to manufacture.

Bioweapons can be deployed by direct contamination (e.g., poisoning of food or water), vectors (e.g., mosquitoes, fleas, or mailed letters), and aerosols (e.g., sprays or bombs). Of the three, aerosols are generally considered the most efficacious means of dissemination. The poisoning of food and the disruption of agriculture are other possibilities.

RADIOLOGICAL AND NUCLEAR WEAPONS After biological weapons, nuclear devices in the hands of terrorists pose the next-greatest threat of mass destruction. In weighing these two types of weaponry, nuclear devices are probably more difficult for terrorists to obtain than biological agents, and a small nuclear explosion would likely inflict fewer casualties than a well-orchestrated, multisite biological attack.

Curiously, nuclear weapons may be a bigger danger now than they were during the Cold War. Unlike nuclear-tipped missile strikes, today's nuclear devices could easily enter a country undetected. Some nuclear bombs are as small as a suitcase or knapsack; others could be concealed in nondescript shipping crates sent via ship from overseas by persons unknown.

Then, there are "dirty bombs." Here, highly radioactive material, such as uranium or plutonium, is wrapped around a conventional explosive (e.g.,

TNT). The radioactive material might be weapons-grade or could come from spent nuclear-reactor fuel rods or even hospital waste (i.e., materials used in radiation therapy). When the device explodes, the radioactive material is blasted into the air and strewn across the neighboring landscape. A dirty bomb doesn't cause an atomic chain reaction like a nuclear warhead. Instead, it merely scatters radioactive debris, much as a hand grenade sends out shrapnel. In fact, a radiological weapon needn't explode at all. Radioactive powder could be introduced, say, into a building's air-conditioning system.

INFRASTRUCTURE ATTACKS   Analogous to a dirty bomb would be a terrorist attack on a nuclear power plant. Although these reactors are encased in protective shells, it's possible that a large plane loaded with explosives could penetrate a plant and release radioactive material into the atmosphere. Terrorists might otherwise somehow enter a nuclear power plant and disable its safety mechanisms. This could cause a nuclear meltdown like the one that occurred at Chernobyl in Ukraine in 1986, which killed more than 30 people immediately and disbursed radiation over a 20-mile radius, necessitating the evacuation of 135,00 people. At Chernobyl, though, an explosion had rocked the plant, destroying its protective covering, which allowed radioactive material to escape. Ukraine's Ministry of Health now says that 125,000 people have died and 3.5 million others have become ill as a result of the Chernobyl accident.

CYBERASSAULTS   Cybercrime, such as fraud and identity theft, is a growing problem, and cyberterrorism specifically has been called "the threat of the new millennium." That's because today's advanced economies are information-dependent, and most information and data now travel via the Internet. Cyberspace thus represents an enticing target for both criminals and terrorists, as well as hackers, virus-writers, and other malcontents.

The possibilities are mind-boggling. Noting that "the next generation of terrorists will grow up in a digital world, with ever more powerful and easy-to-use hacking tools at their disposal," one computer expert has theorized that terrorists might, for instance, target robots used in Internet-enabled telesurgery to cause serious harm to patients. An alarm was sounded in late 2000 when a hacker used the Internet to break into the University of Washington Medical Center's computerized database. The hacker downloaded confidential information on thousands of patients—and learned a lot more about these people than just their blood types.

"Medical information contains your Social Security number, date of birth, and even a physical description," explained Betsy Broder, who tracks identity theft for the Federal Trade Commission. "All of those are keys that people could use to exploit someone's financial identity, as well as their personal identity." *USA Today* reporter Greg Farrell, reporting on the University of Washington incident, added, "The most pernicious use of medical information involves the elderly or extremely ill. If an identity thief knows an individual is close to death, he or she could take out a life insurance policy in the victim's name, naming themselves or an accomplice as the beneficiary."

IDENTITY THEFT Identity thieves operate in ingenious ways. An impostor may call a credit card issuer pretending to be someone else and ask that an account's mailing address be changed. He'll then run up charges on that person's account. Because the bills are being sent to the new address, it usually takes time before the victim catches on to the scheme. Indeed, identity theft usually goes unnoticed for twelve and a half months.

An identity thief alternatively might open a new credit card account, using another person's name, date of birth, and Social Security number. After the bill goes unpaid, the delinquency is entered into the credit history of the victim, who may remain wholly unaware of the crime until, say, he applies for a loan and is rejected because of bad credit. Impostors have been known even to use another person's name to open bank accounts, take out auto loans, and apply for jobs. An identity thief might go so far as to file for bankruptcy *under another person's name.* That way he can avoid paying debts he has incurred in the other person's name or forestall eviction from a home rented, again, using someone else's identity.

1

PROTECTING YOURSELF AND YOUR FAMILY

CHAPTER 1

Becoming Security-Minded

Whether we realize it or not, we're all security-minded and safety-conscious to one degree. Most of us lock our doors, bypass crime-ridden neighborhoods, don't unfurl billfolds in public, and stop mail and newspaper deliveries when we're away on vacation. We also look first before we cross streets, drive defensively, and know how to dial 911.

Terrorism prevention and protection means adding a new set of procedures and routines to the list of precautions we already take. It's easier said than done, however. Think of heart-attack victims. If they're lucky enough to survive their first attack, they almost immediately become health- and fitness-conscious. They start taking care of themselves by going on diets, losing weight, and watching what they eat. Some stop smoking and drinking, and many join health clubs and begin exercising regularly. The intriguing question is, how would these heart-attack victims have fared if they had followed this lifestyle in the first place? Many of them would, most probably, have averted the early onset of heart disease.

Our personal health, safety, and security, in other words, begin in the mind. We want to build concentric rings of defense, which increase your security and safety by making it harder for terrorists to victimize you. Think of the rings as encompassing both the locations at which you spend a good deal of time and the activities you engage in that could bring you in contact with terrorists. Then, it's a matter of instituting security measures and changes in routines that will lessen your chances of falling prey to terror.

Know the five means of staying safe.

In general, there are five ways of staying safe: avoidance, prevention, escape, assistance, and time. All of your preplanning, your thinking, and your strategies should revolve around those five means of survival. Avoidance means not being in the wrong place at the wrong time. Prevention is the erection of barriers that protect you and your possessions. Escape is the recognition that you don't have to take your fate lying down. Assistance requires knowledge of how and when to seek outside help. And time is that precious commodity that keeps you alive long enough for help to arrive.

> ✔ If you frequently travel around town alone, particularly at night, invest in a personal attack alarm or, at the very least, a loud-sounding whistle.

Stay away from crowds.

That's the single best piece of advice that anyone seeking to avoid terrorism can follow. Terrorists most often strike at crowded locations and events, such as shopping malls, large department stores, sporting events, discotheques, historic landmarks, tourist spots, airports, commuter hubs, movie theaters, fashionable restaurants, bars, and trendy resorts. By avoiding these events and locations, you avoid the terrorists.

> Dark, deserted streets are other places to avoid.

Think of security as a lock that buys you time.

Security, like a lock, can never be 100 percent perfect. It's not meant to be. The purpose of security measures is to buy time to escape, get assistance, or wait for help to arrive. Once you drop the fictitious ideal of perfection, your imagination will be free to conceive of a security plan that's unique to you and your loved ones.

Adapt your planning to your circumstances.

Everyone's circumstances are different, so adapt security measures to your situation. Put each contemplated security step in context. Ask yourself, why am I doing this? The answer should be—survival. So, again, engineer all of the security measures you take with the goals of avoidance, prevention, escape, getting help, or buying time.

Get informed about potential terrorist targets in your area.

Begin by looking at a map and drawing radii of 10 to 20 miles around your home and workplace. Then, investigate the potential terrorist targets within those zones. Ask your electric utility the location, if any, of the nearest nuclear power plant and repositories of spent nuclear fuel. Ask your local emergency management office, including police and fire departments, if other critical infrastructure sites (e.g., dams, chemical plants, telephone switching centers, etc.), hazardous or radioactive waste dumps, or major industrial complexes handling hazardous materials are in the vicinity. Ask if any local or state emergency plans are available to the public.

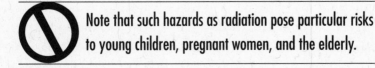

✔ Know how to shut off the gas, water, and electricity in the event utility lines are damaged in a major attack (or in an earthquake).

🚫 Note that such hazards as radiation pose particular risks to young children, pregnant women, and the elderly.

Map assorted evacuation routes.

Most local authorities have contingency plans for the evacuation of nearby residents and workers in the event, say, of an emergency at a hazardous chemicals facility or a nuclear power plant. Traffic flow would, supposedly, be directed away from the site. However, at least half the residents and workers within 10 miles of the facility would probably flee, creating chaos on the roads—particularly on highways and other major thoroughfares. So, ask local emergency managers for copies of official evacuation arteries and then review your road maps in search of alternative avenues of evacuation, routes likely to be less traveled in an emergency.

✔ Weigh the option of staying put in such an emergency. Road rage, traffic jams, and outdoor exposure to toxic materials may pose more danger than simply staying in your home or workplace. Stay tuned to your local television and radio emergency stations for official information and instructions in a crisis.

Don't return home with a fuel tank that's less than half full.

In an emergency evacuation because of, say, a chemical or nuclear attack, you may have to travel a long way to reach safety. The roads will, of course, be jammed, and traffic may crawl. The last thing you want to happen is to run out of gas. So, keep your fuel tank at least half full.

Gassing up before coming home will serve another, perhaps more important purpose: it will test your antiterrorism vigilance. Everyone has a tendency to let his guard down in times of peace and tranquillity. Terrorists count on that. It may indeed be one reason why they often let considerable time elapse between attacks. Your gas gauge is an easy way to tell whether you've been lulled into a false sense of security—or just gotten lazy.

Never carry gasoline cans in your vehicles, for the danger of a fiery crash is worse than running out of fuel.

If you have young children, find out whether their schools have evacuation plans in place.

Local authorities have plans for the transportation of students, as well as children in day care, the hospitalized, and nursing home residents, in an emergency. Speak with school officials or teachers about the plans affecting your children and ask what steps you as a parent should take in an emergency evacuation of local schools. Find out, in particular, if students will be taken to a predetermined mass-care facility and whether you should plan on going there, too, in an emergency.

Be alert to government warnings.

The latest U.S. government terrorism warnings can be found at the Office of Homeland Security (www.whitehouse.gov/homeland/), which also has links to state offices, the Centers for Disease Control and Prevention (www .cdc.gov), the Central Intelligence Agency (www.cia.gov), Defense Department (www.defenselink.mil), Federal Bureau of Investigation (www.fbi.gov), FirstGov (www.firstgov.gov), and the State Department (www.state.gov), which offers terror-related travel advisories.

✔ Buy an all-hazards alert radio, which automatically provides not only warnings of weather hazards like hurricanes and tornadoes but also emergency messages from state and local authorities.

Treat official reassurances circumspectly.

In attempting to calm the public's nerves and cover their own backsides, many government officials at times of crisis have offered unreliable reassurances about public health and safety. Look at how the anthrax outbreak was so mishandled at the start. Why, for instance, were congressional office buildings closed for decontamination but not the U.S. Postal Service facilities that sorted and routed the tainted mail to Capitol Hill? Clearly, some public-health officials were out of their depth. They offered advice that may have cost people their lives. So, take any reassurances made by government officials with a grain of salt. Be circumspect. If you find their reassurances hard to believe, don't believe them.

✔ Stay informed, but don't rely solely on television for the latest news on terrorism and other emergencies. Make it a habit to read local and national newspapers and check out online news sources. Many of these are listed in the reference section at the end of the book.

Don't become complacent, and don't let your guard down.

Immediately following the attacks of September 11 and the discovery of anthrax in the mail, everyone everywhere became anxious about personal safety and the safety of loved ones. As time went on and memories faded, people started getting "back to normal" in terms of commuting to work, shopping, traveling, and handling mail. It was important that people did this, because it showed that they wouldn't be cowed by terrorism.

There's a downside to "normal," however, and that is that people can become complacent. The danger of terrorism is real, and that's something people all over the world mustn't forget. The defeat of Al Qaeda and the Taliban in Afghanistan haven't made the threat go away. Sixty or more countries have terror organizations and cells with global reach. This means, of course, that people shouldn't let their guard down, lest they become the next victims.

HOW TO ASSESS YOUR RISK

When it comes to crime or terrorism, most people reckon that there's little they can do to avoid becoming a victim. Oh, sure, you can know enough to avoid dark alleys late at night and not to flash expensive jewelry or carry wads of cash in crime-ridden neighborhoods. But beyond these lessons in Crime Prevention 101, most people are resigned to the possibility of becoming a victim.

Truth is, things can be done to lower your risk of violent harm. The process begins with a self-assessment of risk.

To gauge an individual's risk level, we can take a lead from the techniques of statistical analysis. Data and information can be sorted in as many as five different ways—by the alphabet, time, category, hierarchy, and location. If we try to apply this methodology to personal safety assessment, we find that four of the five techniques work—namely, time, category, hierarchy, and location. Only one method, using the alphabet, isn't applicable (unless someone decides to pick their targets out of the phone book and start with the letter *A*).

We know, for instance, criminals often choose tightly congested sidewalks and malls, which offer a large number of potential victims and good escape routes. Others prefer to come out after dark, because it ups their chances of successfully perpetrating a crime and not getting caught. Terrorists similarly favor rush-hour attacks, when the high density of worker traffic increases the damage done by bombs and the like.

Here's how to go about the sorting process. And don't worry if you find overlaps; they're to be expected.

Ask yourself whether you work or live in a geographic location, or frequent places, likely to be targeted by terrorists.

The first location to consider is the country in which you reside. Unless you live on some remote island, you can probably consider yourself a potential target from the start. Then, look at the history of terrorism in your part of the country. In the United States, for example, the Northeast has experienced twice as many incidents of terror—by both foreign and domestic terrorists— as has the South.

Next, look more specifically at the city (or cities) in which you live and work or travel to often. Is it a large, internationally known city? Is the name of your city synonymous with "America," say, or "Canada"? Is your city an important center of government, finance, or business? Is it located near an ocean port? Now be even more specific. Is the street on which you live or work internationally recognized? Is your neighborhood normally jammed with people?

The pattern has been for terrorists to go after either high-profile targets or places where lots of people congregate. The World Trade Center, for example, was attacked because it represented Wall Street and American capitalism. London has been a frequent target of Irish republican terrorism because it's the capital of the United Kingdom. And London's Oxford Street has been bombed because it's a main shopping thoroughfare.

This isn't to say that terrorists won't change their habits. They might eventually find it easier to conduct attacks in suburban towns or other out-of-the-way locales where police surveillance is comparatively lax and the public is less inclined to believe that terrorists would ever strike there.

Categorize the places you live, work, and frequent or stay while traveling.

It's easy to confuse category with location, but the two aren't the same (although they may overlap). A category site represents a tempting target no matter where it's situated. Many categories of targets have symbolic importance. Others are merely densely populated. It could even be that an event, such as dedication or award ceremony, is the target.

Take, for example, the Olympic Games. In 1996, a bomb detonated in Centennial Olympic Park in Atlanta, Georgia, where around 70,000 people were celebrating the Summer Olympics. One woman died, and 110 people were wounded. It likely wasn't important that the event was held in Atlanta, for what probably attracted the bomber were both the symbolism of the Olympics and the large crowds associated with the games. At the Munich Olympics in 1972, 8 Palestinian Black September terrorists seized 11 Israeli athletes. A rescue attempt by West German authorities left nine hostages and five terrorists dead.

Other categories are also based a high concentration of people, such as shopping malls, big department stores, and entertainment spots (e.g., restaurants, sports stadiums, discos, and movie theaters), as well as airports, rail stations, subways, and bus terminals. Parades, outdoor concerts, and large rallies are other examples. Fame, too, can be a category, as in luxury apartments, swanky hotels, and national symbols.

Still other categories include government buildings; headquarters of large, internationally known corporations; sites reflective of key industries (e.g., oil refineries and chemical plants); infrastructure sites (e.g., electric power plants and telephone switching centers); and seaports. Do you, for instance, live near a facility like a nuclear reactor or chemical plant, which, if attacked by terrorists, could pollute its surroundings with radioactive materials or toxic fumes? Do you work near a shipping facility that handles containers from overseas that could potentially contain a hidden nuclear device? Does your job regularly take you to federal courthouses or other government facilities?

Yet another category involves your source of drinking water. Do you have a well, or do you get water from a municipal service? Although the danger from a poisoned reservoir is low (because the contaminant would be too dispersed and would deteriorate rapidly), a pump station could conceivably become a target of chemical or biological terrorism.

Review your daily schedule and other routines to see what times are associated with typical terror attacks.

Do you commute to work or school almost invariably at rush hour? Similarly, do you have a transportation-related job, such as a subway police officer, train conductor, bus driver, or rail ticket agent, that places you near large numbers of commuters at rush hour? Terrorists frequently do their dirty work at rush hours for the obvious reason that a bomb or other device detonated at such times is likely to kill or injure the largest number of people possible. The nerve-gas attack on the Tokyo subway in 1995 took place at the height of the morning commute. Bombings of commuter trains and subways in France and Britain, too, have been timed for rush hours.

Other time-specific terrorist incidents might take place at large holiday gatherings (e.g., Fourth of July fireworks displays, tree- or menorah-lighting ceremonies, and New Year's celebrations), on Election Days, or on peak travel days (e.g., immediately before and after Thanksgiving and Christmas). Holiday sales at department stores are another time-related example. Yet other

Risk Assessment Scorecard

Points for each item are in parentheses. Allot yourself points on all applicable lines; multiply where required, such as your hotel stays per year. Add up all your points and compare your score to the risk levels cited at the bottom.

1. LOCATION

Work in:
Large city (10) _____
Medium/small city (3) _____
Town/rural (1) _____
Landmark district (7) _____
Financial district (6) _____
Shopping district (6) _____
Downtown (4) _____
Seaport/airport (8) _____

Live in:
Large city (10) _____
Medium/small city (3) _____
Town/rural (1) _____
Landmark district (7) _____
Financial district (6) _____
Shopping district (5) _____
Downtown (4) _____

2. CATEGORY

Live in:
Building with public
    parking (6) _____
Landmark building (5) _____
Tall urban building (4) _____

Work in:
Government building (10) _____
Landmark building (9) _____

Work in (continued):
Building with public
    parking (9) _____
Tall urban building (8) _____
Transportation (8) _____
Large store/mall (6) _____
Restaurant/hotel (5) _____
Facility open to public (5) _____
Small urban store (2) _____

Work for:
Law enforcement/
emergency unit (30) _____
Transportation (15) _____
Government (10) _____
Retailer/restaurant/hotel (7) _____
Infrastructure
    (e.g., electric power) (8) _____
Famous company (7) _____
Private security (7) _____
Vital industry
    (e.g., oil, computer) (6) _____
Education/health (2) _____

Commutation:
Take mass transit (6) _____
Drive to work (1) _____
Use major train/bus/air
    terminal (15) _____
Commute via major
    bridge/ tunnel (6) _____

Subtotal _____

Work as:
Police/military/emergency
   responder (30) _____
Traveling salesman (10) _____
Senior executive (9) _____
Government worker (8) _____
Transportation worker (7) _____
Infrastructure worker (6) _____
Private security (6) _____
Retail store worker (5) _____
Hotel/restaurant worker (5) _____
Middle manager (4) _____
Office worker (3) _____
Factory worker (2) _____
Healthcare worker (2) _____

Travel/entertainment:
Domestic flights
   (1 per trip/yr) _____
Travel abroad
   (4 per trip/yr) _____
Long distance train/bus
   (1 per trip/yr) _____
Hotel stays
   (1 per stay/yr) _____
Visit historic sites
   (1 per visit/yr) _____
Sporting events/concerts
   (1 per event/yr) _____
Regular moviegoer (4) _____
Dine out often (4) _____

Shopping:
Frequent mall/department
   store shopper (6) _____
Small store shopper (1) _____
Heavy shopper (6) _____
Moderate shopper (3) _____

3. TIME

Work 9 to 5 (7) _____
Work off-hours (3) _____
Large city commute:
   Rush hour (20) _____
   Non-rush hour (10) _____
Other commute:
   Rush hour (5) _____
   Non-rush hour (2) _____
Travel on major holidays
   (2 per trip/yr) _____

4. HIERARCHY

High crime area:
   Work in (20) _____
   Live in (20) _____
Area already hit by terrorism
   Work in (15) _____
   Live in (15) _____
Work in law enforcement/
   fire/military (30) _____
Work in transportation (10) _____
Work for government (8) _____
Work in financial hub (7) _____
Travel abroad often (15) _____
Frequent flyer (7) _____

     Subtotal _____

     TOTAL _____

Personal Risk Assessment

Low Risk: Below 50
Moderate Risk: 50–150
High Risk: Above 150

specifically timed attacks could relate to terrorism anniversaries of one sort or another—say, the date September 11 in the years to come, or August 7, the date of the 1998 U.S. embassy bombings in Africa.

And don't forget the time factor when it comes to matters of everyday violent crime. Do you often travel alone at night along deserted streets or wait at empty stations for infrequent trains? There's no sense in protecting yourself from terrorism only to be murdered by a street thug.

Ask yourself where you stand in the hierarchy of terrorist targets.

The single most important question you have to answer is, have other people in your line of work been victimized by terrorism because of what they did for a living? The answer says a lot about your odds of being victimized, too.

Well-known business executives especially need to be concerned as a number of assassinations of chief executives have occurred in Europe over the years. Then, there are also the ever-present worries about kidnappings, extortion, and ransom demands. But you needn't be a celebrity to be listed as a target. It could be that the widely recognized company you work for is the name that the terrorists prize. You might just be your firm's unlucky representative.

Employees also need to give the following questions some thought: Do you handle mail routinely? Do you work in the lobby of your building or near windows exposed to the street? Do you work at a critical infrastructure site, such as a power plant? Do you work for a major industry of national significance, such as oil or computers? One infamous bombing by the FALN, the violent Puerto Rican nationalist group, took place in 1978 at the headquarters of the Mobil Oil Company in midtown Manhattan; the device exploded in the street-level employment office, killing one man and injuring several other passersby on the sidewalk.

Other hierarchical employment profiles include workers at facilities handling large volumes of people (e.g., airports, major subway stations, or railheads), sports arenas, and large retail stores. Then, too, there are those at greatest risk from hijackings, such as pilots and flight attendants, or assassination, such as law enforcement. Cargo handlers at airports and seaports are similarly at risk from bombs and other concealed devices.

Hierarchical selection also applies to geographic location and travel habits: Do you live or work in a city high on the terrorist list? Do you often visit those cities on business or for pleasure? Do you travel often, particularly by air? Do you go overseas frequently, notably to countries where the incidence of terrorism is high?

Put two and two together and see how you can change your ways so as to reduce your exposure to terrorism.

In making this personal risk assessment, you want to see how exposed you and members of your family are to possible terrorist assault. The more slots you fit into, the higher the chances are of your becoming a victim. If you find a lot of overlap, say in terms of the city you work in and the time of your daily commute, you need to be especially concerned. What you want to do next is try to lessen your risks by changing times, routines, habits, locations, and so on—and perhaps even moving or switching jobs. Instead of being in the wrong place at the wrong time, simply don't be there. A bridge might be destroyed at rush hour by a truck bomb, but you don't have to be on it. A suicide bomber could kill dozens of holiday shoppers at your favorite store, but you don't have to be among them. Subway riders on the way home from work may be overcome by deadly gas, but you don't have to be on that train. You get the idea.

As a case study, consider the predicament of New Yorkers living above underground public parking garages.

Manhattan is without doubt one of the world's costliest places to live. Housing is not only expensive but can also be difficult to find. Against this backdrop, consider the plight of New Yorkers who reside in buildings with underground public parking facilities. Their predicament presents an ideal example of how to assess personal risk. Manhattan apartments with underground parking are typically among the finest residential buildings in the city. They tend to be modern, well located, and filled with amenities, and they often offer spectacular views. But living above a public parking facility in a city that has many times been a target of terrorism presents a serious problem, especially because underground garages have been used before by terrorists to plant truck and car bombs in expectation of destroying the structures above. Residents of such New York apartment buildings have only three clear options: (1) move; (2) get the parking garage closed to the public (and also install security doors and video surveillance to ensure that only tenants can get access, although a terrorist could conceivably rent or buy an apartment in the building and gain entry to the garage that way); (3) do nothing but cross their fingers. We leave it to you to draw your own conclusion.

BASIC SAFETY PRECAUTIONS

To build, in effect, a perimeter defense around your home, family, money, and possessions, think in terms of concentric rings—an outer ring encompassing a smaller ring, then another smaller ring, and so on.

Place you and your family at the center of this mini–solar system. Look at the various ways you interact with your community and your region. Take note of the methods of interaction, such as driving, walking, shopping, etc. Then identify those places in which you spend most of your time, such as home, school, and work. Next see how the outside world comes in contact with you. The mail, the Internet, and financial transactions are among the many ways. Finally, look at your basic necessities, like food and water.

Once you've identified these concentric rings, you can begin to develop and implement various means of protecting yourself and your family that are specifically geared to address your biggest vulnerabilities.

Safeguarding your home against terrorism and crime begins with locks and lighting.

The same security procedures will help to protect you and your home from both terrorists and criminals. The most important of these are sound locks and good lighting. Install tumblers that can't be picked (at least not easily), and reinforce doorframes so doors cannot be forced or pried open. Install peepholes in windowless exterior doors.

On basement and lower-floor windows, be sure the locks are in working order and install jimmy-proof devices. However, if you install window gates or locks, nail the key for each one on the wall near the window—far enough away so a burglar couldn't reach it from the outside but near enough so you could find it quickly in an emergency. Use horizontal "charley bars" on sliding doors. People often use wooden sticks to secure sliding doors; burglars lift those out using coat hangers. Upgrade your garage-door locks. Criminals in passing cars using various electronic devices can open many garage doors. Have an accredited professional fit a tamperproof lock.

Install outdoor lighting connected to motion detectors, and keep hedges and bushes trimmed. Have some lamps hooked up to timers to go on and off when you're away overnight. Consider subscribing to a professional security service that will call the police or fire department in the event of a trespassing or fire emergency. And if you're so inclined, install a video security system of minicameras on your property to watch and record all comings and goings.

> ✔ Be sure the address of your home is visible from the street. It makes it easier for emergency services to find you, especially at night.

> 🚫 Never leave spare keys hidden outside the house; instead, add a house key to your ring of car ignition and door keys. Remember, though, when giving a parking attendant your car key, give him only the one to the ignition.

Be attuned to anything out of the ordinary in your neighborhood.

There's a fine line between being a conscientious neighbor and being nosy. Still, in this day and age, it pays to be aware of your surroundings. Try to develop a rapport with your neighbors so you may jointly work to safeguard the neighborhood. If you hear or spot something unusual, wait to see if anything develops. If you suspect a crime is being committed, inform law enforcement. Your actions could save a life.

✔ Familiarize yourself with the locations of police and fire stations, hospitals, public telephones, and 24-hour stores and eateries near where you live and work. That knowledge could come in handy in an emergency.

Install effective smoke and carbon-monoxide alarms.

Make sure you have an adequate number of smoke alarms in your home (at least one on each floor), including ones that can detect any carbon monoxide that might escape from your furnace or garage. Test the smoke detectors routinely, and replace their batteries regularly. It's said that smoke detectors more than double the chance of fire survival.

✔ To ensure your smoke-alarm batteries are fresh, change them in the spring or fall when you reset your clocks, or on an annual national or religious holiday.

🚫 Smoke alarms only have a useful life of 10 years, after which they should be replaced even if they still appear to work.

Have fire extinguishers around the house.

Have fire extinguishers handy and know how to use them; your local fire department likely offers training courses. Recognize that there are different types of extinguishers to combat different types of fires. A-type extinguishers are used to fight common combustible fires involving wood, paper, and cloth. The B-type is used to extinguish flammable liquids, such as gasoline, grease, and oil. And the C-type combats electrical fires in, say, heaters, computers, toasters, and other household appliances. You can buy these different types in various combinations, such as BC or ABC fire extinguishers.

Consider buying collapsible ladders for upper-floor escapes, especially from bedrooms.

 Fire extinguishers require regular inspections to ensure they still function properly.

If you live or work in a high-rise building, practice your means of escape.

Don't just know where the emergency exits are in your building, walk to them every once in a while until it becomes second nature. That way, you won't waste precious time in an emergency. You'll know exactly what to do and where to head. Your expertise could, in fact, prove invaluable to your coworkers and family in the event of an explosion or other emergency.

✔ Should you need to crawl to an exit, make some mental notes beforehand. Develop maps of your workplace and home in your mind, using furniture, equipment, doors, and hallways as landmarks to help guide your way.

Assemble an emergency-supplies kit.

Assemble a kit of emergency supplies and store it near your vehicles in case of a mass evacuation. Your kit should include the following:

- water (three gallons per person)
- nonperishable food (enough for three days)
- warm clothing and sturdy footwear
- blankets (or sleeping bags)
- first-aid kit

- flares and flashlights
- battery-powered radio and cell phone, plus extra batteries
- sanitation and hygiene supplies (e.g., toilet paper, toothbrushes)
- tools for repairs, shovel, ax, broom, rope, wire
- disposable plastic bags
- waterproof tarp (or tent) and waterproof matches
- cooking source (e.g., camp stove or Sterno)
- pet food and litter, if applicable

Bring an extra set of car keys, too. You don't want to get locked out of your car miles from home. Pack a can opener, paper towels, premoistened towelettes, insect spray, sunblock, water purification tablets, and disposable plates, cups, and utensils. Also, carry credit and debit cards with you. You'll need road maps and a compass, too.

✔ Be sure spare tires are in proper working condition.

 Stored water and food should be changed every three to six months. Don't store plastic containers on concrete floors, because chemicals can leach through and spoil the contents. Also, exercise caution when drinking from streams, rivers, lakes, and springs that usually contain bacterial contaminants.

Have a properly outfitted first-aid kit.

Both the American College of Emergency Physicians (www.acep.org) and the American Red Cross (www.redcross.org) have compiled lists of items they believe should be in every first-aid kit. The kit should be kept together in one place, and every mature member of the household should know where it is and how to use each item in it. To carry the kit around, a small tote bag is recommended. Take the same precautions with your first-aid kit that you would with any medicine. Store it out of the reach of children, and only use products with child-safety caps.

The items below will provide you with the necessary tools to handle many medical emergencies. All of the items are available at local pharmacies.

Recommended First-Aid Kit Contents

- Acetaminophen, ibuprofen, and aspirin tablets for headaches, pain, fever, and simple sprains or strains. (Aspirin should not be used for relief of flu symptoms or given to children.)
- Ipecac syrup and activated charcoal for treatment after ingestion of certain poisons. Use only on advice of a poison control center or the emergency department.
- Elastic wraps for wrapping wrist, ankle, knee, and elbow injuries.
- Triangular bandages for wrapping injuries and making an arm sling.
- Scissors with rounded tips.
- Adhesive tape and two-inch gauze for dressing wounds.
- Disposable, instant ice bags for icing injuries and treating high fevers.
- Bandages of assorted sizes for covering minor cuts and scrapes.
- Antibiotic ointment for minor burns, cuts, and scrapes.
- Gauze in rolls and in two-inch and four-inch pads for dressing wounds.
- Bandage closures: quarter-inch and one-inch for taping cut edges together.
- Tweezers to remove small splinters and ticks.
- Safety pins to fasten bandages.
- Tongue depressors to inspect the throat and mouth.
- Thermometer to take temperatures.
- Rubber gloves to protect yourself and reduce the risk of infection when treating open wounds.
- First-aid manual.
- Flashlight and spare batteries.
- List of emergency telephone numbers.

Other items should include eyedrops, antibiotic ointments, diarrhea medication, cough medicine, antihistamines, ear and nose drops, cotton safety swabs, insect repellent, skin disinfectant, sunblock, hydrogen peroxide, and premoistened towelettes. A spray bottle containing a solution of 10 percent laundry bleach and 90 percent water can be used to disinfect objects.

 Don't forget to pack any prescribed medications and extra eyewear.

For help in a case of poisoning, call the American Association of Poison Control Centers (www.aapcc.org) at 800-222-1222. If the victim has collapsed or isn't breathing, dial 911.

If a poison is ingested, seek the advice of a poison control center, doctor, or emergency room before taking any medications or inducing vomiting. A caustic chemical could, for example, destroy additional sensitive tissue if regurgitated.

Check out online first-aid advice regarding terrorist-type attacks.

One of the more interesting cost-free manuals available online is the U.S. Army's *First Aid for Soldiers* at www.vnh.org/FirstAidForSoldiers/fm2111. html. The University of Iowa and the U.S. Navy Bureau of Medicine and Surgery also have a Virtual Naval Hospital online (www.vnh.org) with invaluable information for patients and health providers on what to do in the event of a biological, chemical, or nuclear attack.

Know how to prevent or control shock.

Severely injured persons often go into shock. Knowing how to prevent or control this can be helpful. However, a few important words of caution are in order first. If you suspect the victim has a broken neck or back, don't move him at all. Similarly, don't move him if he has head or abdominal wounds or a broken leg. Simply cover him with a blanket or piece of clothing to help keep him warm. Otherwise, lay the person on his back, unless a sitting position makes his breathing easier. Elevate his feet so they're higher than his heart; this increases blood flow to the brain and other vital organs. Prop up his feet with an object that would easily fall over. Loosen any binding clothing around his neck and waist. Cover him with something to keep him warm. Speak with the victim in a calm and reassuring manner to lessen his anxiety level and blood pressure. Tell him he'll be all right, that he's being cared for, and that help is on the way.

 It's recommended that if a chemical, biological, or radiological device has been used, you not touch victims so as to avoid becoming contaminated yourself.

Be prepared to isolate yourself and your family in the event of a communicable-disease outbreak.

In the event of a terrorist-inspired outbreak of smallpox or other communicable disease, the best step to take to avoid exposure would be to limit contact with others until the authorities give an all clear. Businesses should shut down, and travel of any kind should be eschewed. To be on the safe side, it would be smart to take a few minutes to make a list of all the persons with whom you had been in immediate, personal contact within the previous week or so. If it turns out that you have contracted smallpox, say, that list of names could prove invaluable in saving lives.

Have the necessary supplies to create a safe-haven room.

Inhalation of deadly or incapacitating germs, chemicals, or radiological agents is the greatest danger to civilians in the event of a terrorist attack with a weapon of mass destruction. Be able, therefore, to turn a room in your house, or even an entire small apartment, into a safe haven. Select a large, inner room on an upper floor, preferably with access to a bathroom and a telephone. Don't pick a room that has an air conditioner built into the wall or window. Store plastic/duct tape and sheets of heavy-duty plastic (six mil at least) in or near the room so you can tightly seal all doors and windows in an emergency. Also have cloth and cotton wool to plug any large gaps, including keyholes. If an emergency arises, bring your survival kit of food, water, first-aid kit, portable radio, flashlights, and other necessities into the safe-haven room and seal yourself off until the all clear is given.

✔ If there's a danger of a nuclear explosion, go to the basement of your house or building to avoid the blast, thermal radiation, and radioactive fallout.

🚫 Don't pick a room that faces any nearby city that might be the site of an attack with a weapon of mass destruction, and don't pick a room that faces into the prevailing winds in your area.

🚫 Store-bought bottled water isn't meant to be kept for long periods; the plastic containers tend to fail or leak after about six months. It's thus best to purchase special containers intended for long-term storage or five-gallon water bottles from a private distributor. Never store plastic water bottles directly on top of a concrete floor, because the concrete will leach chemicals into the bottled water and degrade the plastic containers.

Squirrel away a modest amount of cash in case ATMs go down.

Power outages are always a possibility. Millions in the northeastern United States and southern Canada were blacked out in the summer of 2003 because of what should have been only a minor glitch in the system. Image the havoc that terrorists could wreak. And what about cyberattacks on banking institutions?

Any of these scenarios could make cash difficult to come by. Automated teller machines (ATMs) would likely be out of service. Credit and debit cards mightn't work at your local stores, and electric-powered cash registers could be frozen shut. Cash would be at a premium until the authorities sorted affairs out. So hide away a few dollars for an emergency.

SPOTTING SUSPICIOUS BEHAVIOR

Historically, when eyewitnesses have been debriefed after a terror incident, they almost invariably agree on one thing: the perpetrator looked suspicious from the start. It takes a highly professional, extremely well-trained terrorist to have the poise needed to avoid detention completely. The pros are as cold as ice and rarely give themselves away. But not all terrorists are that well trained. Most are nervous and overanxious. Others may be clumsy or dim-witted.

Be mindful, too, that many terrorists are engaged in activities other than actual attacks. They may perform ancillary functions, such as reconnaissance, bomb making, and weapons procurement. You need to be on guard, therefore, in case a terrorist surveillance team is targeting the building you work or live in for possible attack.

Recognize the telltale signs of suspicious behavior, including more than just terrorism.

Anyone about to commit a heinous act will often look out of sync with their surroundings. Their behavior will be inconsistent with that of the other people around them. Their focus may appear to be elsewhere. They won't act like everyone else or share the same interest or concern (e.g., the bus is late, the line is moving too slowly, etc.). They will be nervous, perhaps jumpy, because they know what they're about to do. Something about them will seem strange or unusual. A group, for instance, may exit the same car or taxi but then act as if they don't know each other. That, you know, is odd. Or, a group may enter a facility together and then intentionally take seats far apart from one another. That, too, is strange. By now, you can probably tell something's going on. Your job, as a good citizen, is to alert law enforcement or building security.

> ✓ On occasion, terrorists will give themselves away by the clothing they wear. They may have a winter coat in the summertime or be overdressed at an event that calls for casual wear.

> ✓ Don't assume that a terrorist has to be a man. Women, too, participate in terror attacks, conduct pre-attack surveillance, and carry out suicide bombings.

Be aware of your surroundings at all times and alert to trouble.

Surprise is a hallmark of terrorism. However, the nature of the terrorists' activities is such that they can't help but display some signs of what they intend. It may be the odd handling of a parcel or suitcase. They may be carrying guns or assault weapons that cannot be concealed completely. They may spill the beans in communicating too loudly with one another or talking on the phone. So stay alert and take in your surroundings with watchful eyes and open ears.

Wherever you are, scan the premises for the emergency exits and staircases, and think about potential escape routes. Make this a habit, especially when you're in a facility for the first time.

🚫 Take special care when you're near heavy objects that could fall on you or glass that could shatter in an explosion.

Look to see whether the terrorist is wearing a facial disguise.

If a suspected terrorist has his face covered in a disguise, it signals that he wants to stay alive and plans to escape before, say, a bomb goes off. This means, of course, that you, too, will have time to escape, so exit the scene promptly and contact the authorities.

Notice someone deviating from standard operating procedure.

We usually know, at least subconsciously, when someone is acting out of the ordinary. We get used to seeing certain routines performed in much the same way by various people functioning in the same capacity. We know, for instance, how janitors go about their chores, or repairmen or security personnel. If you spot a deviation from the norm, you may have spotted an impostor. Inform the authorities of your suspicions.

Determine whether someone is paying too much attention to one particular thing.

A terrorist about to commit an act of violence knows his intended target. He'll, therefore, tend to fixate on it. Yes, he may avert his eyes every once in a while so as not to appear too obvious. Still, he can't help himself. His target is there, right before his eyes; he must look at it. He may have waited and trained for months or years for this moment. His concentration on his target will be immense. From the perspective of an outside observer, his fixation will seem odd and unusual. It will appear to be too much. So, again, contact the authorities to inform them of your suspicions.

✔ A terrorist will often act oddly around police or security officers. He will pay a lot of attention to them from a distance, but when an officer draws near, he will put on an act, feigning ignorance and appearing oblivious to the proximity of the officer.

Take note of anyone trying to gain access to off-limits areas and security doors.

Pay attention to anyone attempting to enter a secure area or specially locked door without the proper uniform, visible credentials, or functioning keys or push-button codes. Raise an alarm if the person is having difficulty opening a security door, or if you see him leave and return to the same door and still be unable to open it.

Be attentive to deliberate acts of concealment.

Carrying concealed weapons and bombs can be difficult, and terrorists sometimes give themselves away through clumsiness or overt attempts at concealment. Watch for persons who may appear to be hiding something under their jacket or a coat hung over their arm. The bundle of a pistol strapped above an ankle, tucked into a waistband, or held in a shoulder holster often can be seen, especially when the person moves. Also, be on the lookout for anyone who is overly protective of a seemingly innocuous package.

Be alert to target reconnaissance.

Well before an attack, terrorists will conduct detailed surveillance of a target site. At first, it may be only to determine whether a site would make a good target. After a site has been chosen, additional reconnaissance will be carried out. Signs include photographing, videotaping, and mapping the site, inquiries about security staffing and procedures. Terrorists will also be interested in locating a facility's entry and exit points, as well as gathering information on adjoining streets, major roadways, and pedestrian traffic. They may even try to determine room sizes and heights, plus the thickness of walls. One of the attackers may conduct a personal walk-through of the facility, testing security and getting a firsthand impression of the target.

Keep an eye out for unattended packages.

Unattended parcels, boxes, briefcases, or luggage left in a public place should assiduously be avoided. Terrorists not wishing to blow themselves up in a suicide bombing use innocent-looking items to plant bombs containing timing fuses or remote-controlled detonators. Years ago, the head of a German bank was killed when his car passed next to a bomb-laden bicycle that had been left on the roadside. If you see an unattended item, particularly at an airport, mass-transit station or terminal, or a crowed street or store, alert the police immediately. Don't touch it or move it. It may turn out to be nothing. But these days, you can't assume anything is harmless. Even flashlights have been turned into bombs.

 Bombers tend to handle their deadly devices with care, so watch for anyone gingerly placing a package or other item in an unusual spot.

🚫 Alert security or the police if you see someone put a package at the back of a store shelf, in a planter, behind a curtain, or in some other hiding place.

Pay attention when a suspicious person drops an item into a curbside mailbox or garbage receptacle.

Terrorists have been know to drop package bombs into street containers, such as mailboxes and garbage cans, then flee before the explosion. Pay attention if someone arouses your suspicions. If you see him drop something into a street container and hasten off, move away. Alert the people around you to flee, and contact law enforcement.

 Be alert, too, to a passing vehicle that stops to deposit an item on the street or in a container and then speeds off.

Know the warning signs of truck bombings in progress.

The FBI, though the National Infrastructure Protection Center (NIPC, www.nipc.gov), asked Americans in March 2002 to be alert for signs indicating plans to construct and detonate truck bombs. The agency had conducted an analysis of truck bombings to "determine whether any unique characteristics exist that might help identify, in advance, potential terrorist activity." It found that terrorist attempts might be preempted by remaining alert for a number of "indicators," outlined below. "While the presence of an indicator does not in and of itself suggest terrorism as a motive," it said, "the FBI's analysis reflects that further examination of the particular circumstances of each case might be in order when one or more of the following indicators is present." The FBI's truck-bombing indicators are:

- Theft or purchase of chemicals, blasting caps, and/or fuses for explosives.
- Theft or purchase of respirators and chemical mixing devices.
- Rental of storage space for chemicals, hydrogen bottles, etc.
- Delivery of chemicals to storage facilities.
- Theft or purchase of trucks or vans with a minimum 2,000-pound capacity.
- Trucks or vans that have been modified to handle heavier loads.

- Chemical fires, toxic odors, or brightly colored stains in apartments, hotel rooms, or self-storage units.
- Small test explosions in rural or wooded areas.
- Hospital reports of missing hands or fingers or of chemical burns on hands or arms.
- Chemical burns or severed hands or fingers that have gone untreated.
- Physical surveillance of potential targets (surveillance may include videotaping, particularly focusing on access points).
- "Dry runs" of routes to identify any speed traps, road hazards, or bridges and overpasses with clearance levels too low to accommodate the truck.
- Purchase of, or eliciting access to, facility blueprints.

The FBI encourages individuals to report information concerning criminal or terrorist activity to their local FBI office at www.fbi.gov/contact/fo/fo.htm or other appropriate authorities. Individuals may report incidents online at www.nipc.gov/incident/cirr.htm and can reach the NIPC Watch and Warning Unit at 202-323-3205, 888-585-9078, or nipc.watch@fbi.gov.

Look to see if a stranger is lurking near your parked car.

A terrorist (or carjacker) may try to overtake you as you enter your vehicle. So as you approach your parked vehicle, look to see if anyone is lingering about or hiding behind a vehicle. If your instincts tell you something's unusual, walk away. Either return later, checking to see whether the person has left, or get help from law enforcement or building security. Be especially vigilant when returning to a vehicle left in a parking lot at an airport, bus or rail station, or large shopping center.

✔ If you use a remote-control device to unlock your vehicle's doors, set it so that it only opens the driver's door. Carjackers have been known to enter vehicles when the passenger doors are opened remotely.

✔ Think about your location. In recent years, about 4 in 10 carjackings took place near airports, bus terminals, train stations, and other open areas, while about one in four occurred near the vehicle owner's home or a friend's home. One in five carjackings happened in commercial places and parking lots.

Keep in mind that a terrorist may be one of your customers.

The FBI issued a warning to law enforcement agencies in February 2002 about Valentine's teddy-bear bombs after receiving a tip that a man had purchased several stuffed bears, propane canisters, and BB pellets at a retail store in California. The incident points out terrorists may buy many of their bomb-making materials (e.g., watches, timers, batteries, wiring, etc.) over the counter, so store clerks should be watchful for suspicious purchasers. Before launching an attack, moreover, terrorists will acquire information on their target, perhaps taking photographs and making detailed notes. It's possible that a photo-processing store or computer repair shop might come across this information, in which case law enforcement should be informed. The same goes for dry cleaners and the like, who may come across revealing information left in an item of clothing. In addition, telephone repairmen, cable or satellite television installers, meter readers, or deliverymen might see something suspicious in a home or apartment they visit.

✔ Employees of restaurants, motels, hotels, and bars should alert authorities if they overhear a discussion of a terror plot or if someone boasts of something "big" happening.

 Be especially careful of anyone making requests that seem aimed at acquiring information about security personnel and practices.

Be wary of odd-acting neighbors.

Terrorists residing in a targeted country have to live somewhere until they execute their action. It may be a house, apartment, or hotel room. In any case, they are someone's neighbors. Moreover, they may have lived at the address for quite a long time, given that terrorist "sleeper" cells have been in place around the world for years awaiting their orders. Be particularly suspicious of groups of relatively young foreign men who routinely gather at a neighboring house, apartment, or hotel room and are clearly looking to see if they're being watched. Do they look behind them before closing the door? Do they peek through windows to see if anyone is outside observing their comings and goings? Is a neighbor overly secretive about his identity or what he does for a living? Does an apparently healthy, young man seem to have no visible means of support? Does he not regularly leave for work or school?

Heed police advice regarding apartment building tenants and rental applicants.

In following procedures for completing tenant applicant background checks, the LAPD advises apartment owners and/or managers:

- Make sure prospective tenants have valid state identification.
- Verify that their vehicles are registered to the prospective tenant.
- Be cautious of prospective tenants using only rental vehicles.
- Take note of prospective tenants who have little or no previous rental history.
- Verify the prospective tenant is able to pay rent.
- Verify employment thoroughly.
- Be cautious of prospective tenants seeking month-to-month or week-to-week rentals.
- Pay attention to prospective or current tenants paying rent in cash or with money orders or third-party checks.
- Be cautious of prospective tenants seeking to rent only ground-floor apartments, unless the request is related to a disability.
- Be cautious of prospective tenants who insist on renting only apartments that cannot be seen by other homes or apartments.

- Take note of prospective tenants who claim to operate nonzoned businesses—e.g., jewelry, industrial art, metalwork, electric repair, chemistry, etc.
- Observe new tenants when they move in to see if they are bringing in any unusual items—e.g., machinery; liquid containers, barrels, buckets, or drums; sacks and bags; compressed-air tanks; large batteries; electrical wire; or fireworks.
- Take note of any boxes carried with extreme caution, as well as any unusual or oddly sized packages being brought into units.
- Monitor apartment units for unusual odors, such as gasoline, diesel fuel, ammonia, sulfur, acids, or fireworks.
- Watch for any unusual digging or trenching near ground-floor apartment units.
- Be cautious of tenants who use an apartment in unusual ways—e.g., rarely occupy the unit or overoccupy the unit with more residents than allowed on the rental agreement.
- Look out for unusual guest traffic to a single apartment unit.
- Watch for tenants or guests using special knocks or signal devices, such as hanging towels, open or closed curtains, or a cushion or some other item positioned in a special way.
- Take note if a tenant changes door locks without approval.

Terrorists are most likely to seek housing in newer apartment complexes where tenants are less likely to know each other, high-rise apartment buildings with large numbers of units, or in densely developed areas, the LAPD says. It, therefore, suggests installing surveillance equipment and paying attention to activity in public areas.

Trust your instincts.

For all the importance placed on human intellect, the instincts we're born with also serve a vital function in helping to keep us safe. If someone or something doesn't seem right, pay attention to what your instincts are telling you. Remove yourself from the situation, and if you feel strongly enough about the potential for terrorism or other violence, tell a police officer or security staff of your concern.

Don't hold back from contacting authorities; hesitation could cost lives.

Some people don't like to get involved in anything other than their personal business. Others are so afraid of being thought silly that they'd never contact authorities about their suspicions. In this age, however, to be a good citizen and responsible neighbor means setting aside any reluctance, second thoughts, or hesitancy about contacting the authorities. Tips about suspicious persons and unusual activities will likely prove invaluable to catching terrorists—hopefully, before they've been able to carry out their dastardly deeds.

DEALING WITH STRESS

Americans, indeed the world, went through a lot on September 11. Witnesses to those events won't soon forget the ghastly, surreal sights or the pain and sadness of seeing so many innocent lives lost in a few short hours. Then, too, there was the fear about opening the mail that developed with news of the lethal anthrax letters sent to members of the media and Congress. New worries are cropping up almost daily—concerns about attacks on nuclear power plants or cyberterrorism disrupting vital infrastructure. It's, therefore, worth taking note of the stress that most us have felt since that September morning and discussing a few healthy coping mechanisms.

Don't be afraid to talk about your feelings and thoughts about terrorism.

In the many interviews with the recovery teams, police, firefighters, and construction crews working at ground zero in New York, most everyone has the same message: to cope, we talk. Keeping emotions bottled up inside isn't healthy. So speak with friends, relatives, coworkers, or professional counselors about your worries about terrorism. You'll probably be surprised to learn that you aren't alone in your feelings and thinking. And that knowledge alone can be a great relief.

Staying physically healthy is important to your emotional well-being, so eat right and exercise regularly.

Getting out of yourself by giving of yourself is a great stress-buster.

Stress develops as we internalize our feelings, sometimes so much so that we become totally egocentric. The more we focus on our problems and our worries, the worse they may become. Indeed, our difficulties could spiral out of control, resulting in severe depression, aggression, or substance abuse. A surefire way of defeating stress is to get out of yourself. That means focusing on others. So lend a helping hand somewhere. Give of yourself. Volunteer to help with a charitable cause, pay more attention to your family and relatives, and do the occasional good deed.

Try yoga-type breathing exercises to relax.

While we don't profess to be expert in yoga, we do know one simple exercise that provides considerable relief from stress. What's particularly nice about it is that it doesn't require you to contort yourself into knots. It involves only breathing. Begin by closing your eyes and mouth and breathe in through your nose, but only fill your lungs by about one-quarter capacity. Count slowly to 10, then exhale slowly through your mouth. But don't exhale any old way. Blow the air out. And blow it out slowly, very slowly, as if you were trying to keep a feather afloat. Next, fill you lungs up a bit more, hold the breath, then blow out the air. Again, take in more air, hold it, and blow. Soon you'll feel as if your lungs are filled from your belly to your shoulders. After 10 to 20 minutes of this, you should feel much more calm and relaxed.

🚫 Novices may hyperventilate. If you feel dizzy, remain seated and breathe in and out of a paper bag or moderate-sized container, recirculating the same air over and over until you feel better. This will return the level of carbon dioxide in your blood to normal and relieve your wooziness. And please, don't perform this breathing exercise while driving.

Get out and enjoy life; socialize and keep your intellect in charge.

The world can be too much for us. Take time to smell the roses, explore the world, meet new people, and perhaps do some of the things you always meant to do but never seemed to find the time. You only live once, as they say. Take time for a bit of self-reappraisal. Write down a few things you'd like to do before you turn old. Then do them. Make good use of your weekends and vacations. Get away from the television and read a good book or listen to music. Take a stroll. Life wasn't intended to be all work. Socialize with others, and don't be ruled by your emotions. Keep your intellect in charge.

Reinforce your sense of security by taking proactive antiterrorism measures.

"Nothing in life can prepare you for the horror of an act of terrorism that robs you of your sense of security and, in some instances, a loved one," notes "Coping After Terrorism: A Guide to Healing and Recovery," published by the U.S. Department of Justice's Office for Victims of Crime. "Recovering from a traumatic event will take a long time and will not be easy." It's vital, therefore, to reinforce your sense of personal safety and security. The reading of this book is a start. Gain strength from the knowledge that you are being proactive in implementing antiterrorism measures. Appreciate your self-reliance, and rest assured that whatever can be done, you have done.

EDUCATING CHILDREN

Terrorism survival requires advance planning. For a family, this means developing a game plan—one that addresses not only terrorism but also other common dangers, including house fires, crime, and neighborhood disasters.

Teach your children to dial 911.

There have been instances, much publicized in the news media, where a young child has dialed 911 and saved a life. Less publicized are the times when a child has called 911 and the police end up scolding the parents for letting the youngster play with the phone. What's a parent to do? We suggest you try role-playing. Have adults and children interchangeably act the parts of victim and rescuer, using an unplugged phone to dial 911. And add a third person to serve as the 911 operator. If you have enough people, you could add as many emergency personnel as you wish. The purpose of the exercise is to teach children not just how to use the telephone in an emergency, but also how to tell when there is an emergency and not to be afraid to seek help. In this way, the seriousness of the call will come across, and the chances are low that your child will ever call 911 on a lark.

 Make sure younger children know their last name and home address.

Post a list of emergency names and numbers.

In a convenient spot, low enough for any member of your household to read, tack up a list of important names and phone numbers. These might be relatives, friends, or next-door neighbors who could be contacted in an emergency. Also include your work numbers. When traveling, add the phone numbers at which you can be reached.

🚫 Children should never let an unfamiliar caller at the
door or on the phone know they are home alone, so
teach them to say something like "Mom can't come to
the phone (or door) right now." Remind children never to speak with
strangers on the street, accept gifts from strangers, or go near a
vehicle if the driver or passenger asks them a question.

Practice home evacuation drills with your children.

Seconds can make the difference between life and death in a fire or explosion, so it's wise to have a home evacuation plan laid out in advance. With your children, go over all the different ways they could safely escape in an emergency. Make sure that each room has at least two ways out. Ensure that all windows, screens, and doors open properly and that no obstructions, such as iron gates or heavy pieces of furniture, would block a speedy exit. Tell your children to stay low to the ground to avoid the toxic smoke and gases that collect near the ceiling in a fire. Have them cover their mouths and noses with handkerchiefs or cloth.

✔ To help young children understand how to escape in an
emergency, draw a floor plan of your home and have
your kids trace various exit routes. Quiz your children regularly on
fire safety and disaster planning.

Educate your kids in fire detection.

Teach your children how to tell if fire is on the other side of a closed door by using the palm of their hand to feel for heat on the door, the doorknob, and the crack around the door. Conduct fire drills at least twice a year—holding one in the daytime and the other at night—and explain the symptoms of carbon-monoxide poisoning (e.g., light-headedness, headaches, dizziness, nausea, vomiting, and fainting).

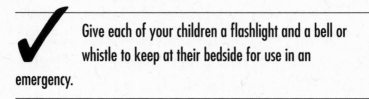

Give each of your children a flashlight and a bell or whistle to keep at their bedside for use in an emergency.

Instruct children never to enter or reenter a burning building.

Make sure your children understand that they are never to enter or go back into a burning building—not to retrieve a pet, a toy, or even you. Explain to them that even if you were trapped inside, you'd want them to remain safely outside, and tell them not to worry, because firefighters will rescue you.

Practice rolling on the floor or ground to put out flames.

This is one exercise kids love. Teach them to drop to the ground and roll if their clothes are on fire. Explain to them that running only fans the flames.

For more information on escape planning and fire education for kids, go to the Federal Emergency Management Agency Web site at www.fema.gov and the U.S. Fire Administration's Web site at www.usfa.fema.gov.

Pick places to gather in an emergency, and ask people to serve as phone contacts.

One of life's worst experiences is not knowing if a loved one is okay following a calamity. To preclude this, pick two places—one near your home and another outside your neighborhood—to meet in an emergency. Also, ask a few friends and relatives to serve as emergency contacts if the members of your family become separated. Be sure you and your children carry those numbers with you. And give your older children mobile phones.

Choose at least one telephone contact from out of state in case the emergency disrupts local phone service or necessitates a large-scale evacuation.

Don't neglect family pets in your emergency planning.

Plan to take your pets in an emergency evacuation. Know, though, that for health and space reasons, some emergency shelters may not accept animals. So have pet carriers that are large enough for your pets to stand up in and turn around. Train your pets to become accustomed to the carriers. If you must leave your pets behind, put them in safe locations, such as bathrooms, with access to flowing water. Don't put them in rooms where hazardous chemicals are stored. Leave them with plenty of dry food that won't spoil, and leave the water running in a sink or bathtub.

Contact your veterinarian to see if he plans to accept animals in an emergency.

Experts recommend separating dogs and cats, because the anxiety of emergency could cause them to act hostilely. Also, separate small pets from large ones.

In a local emergency, consider helping a disabled or elderly neighbor.

If you have to evacuate the neighborhood, besides rounding up the kids, dogs, and cats, you might consider helping a disabled or elderly neighbor.

Talk it over with your family first and then contact the person. It'll do your heart good. At the very least, inform any neighbor who is hearing- or sight-impaired of the local emergency. Inform the police or fire department of the person's location. Do the same for any neighbor who is mentally handi-capped, paralyzed, or unable to walk without assistance.

HELPING CHILDREN COPE

Children can take the news of a terrorist attack (or any disaster) particularly hard. In some cases, a child's reaction can become so severe as to result in post-traumatic stress disorder (PTSD). Here's a brief overview.

Understand that reactions to traumatic events vary with the age of the child.

Children may react to a horrific event almost immediately or long after-ward. They may lose trust in their parents or all adults, fear a reoccurrence of the event, and show signs of emotional disturbance. A lot depends on the age of the child. Children five years of age and younger, says the National Insti-tute of Mental Health (NIMH), may fear being separated from their parents; cry, whimper, or scream; become immobile or aimless; tremble; express fright in their faces; and cling. These young children also may show signs of regres-sive behavior, acting like babies again (e.g., thumb-sucking, bed-wetting, and fearing the dark).

Youngsters 6 to 11 can become withdrawn, disruptive, and irritable. They may refuse to go back to school, display outbursts of anger, and cause fights. They might not be able to pay attention or they may have nightmares. These youngsters may also complain of stomachaches or other imaginary illnesses. They, too, could show regressive behavior or become depressed, anxious, and guilty or even emotionally numb. Adolescents 12 to 17, says the NIMH, may respond in ways similar to adults, suffering flashbacks, nightmares, or emo-tional numbness. Depression could lead to substance abuse. They may dis-play antisocial behavior and decline academically. They could also have difficulty sleeping, complain of phantom illnesses, and feel guilty, angry, and even suicidal.

Talking with your children is a big help.

Children affected by trauma need reassurance, and experts recommend that talking with your children can help a great deal in healing and recovery. Honest answers to questions are the best, though child psychologists emphasize that you should tailor your answers to the age of the child. In general, children want to be assured that they're safe and that their parents are safe. Be open and supportive. Tell them that actions are being taken to catch the people who were responsible for the attack and that the government is trying to make sure it never happens again. Don't, however, promise that it will never happen again. And limit the amount of television exposure your children have to the scenes of violence or disaster.

Listen to your children with love and understanding.

Children should be encouraged to talk following a national trauma. Help them find ways to express themselves, because they need to vent their thoughts and fears, experts say. So be attentive and listen to them with understanding and provide them with reassurance. Assuage their fears, rid them of any extreme worries, and let them know they're loved and being cared for. Recognize, too, that children tend to personalize a traumatic event, expressing concerns about the well-being of relatives and friends. Let them speak or meet with grandparents and other relatives to reassure them that everyone in the family is all right. At the same time, don't dismiss your children's feelings out of hand. Acknowledge their thoughts and appreciate their reasoning. Don't make them feel stupid or think they're wrong to feel as they do. But then explain to them why they needn't feel or think the way they do.

Try to make your children's lives more predictable and stable.

Child psychologists say that parents have to do their part in helping their children overcome the emotional disturbance caused by acts of terror or other disasters and violence. Reintroduce routines into the home to make your children's lives once again predictable and stable. Make it a point to gather together as a family. Play games, engage in sports, read to your children, plan outings, and visit relatives and friends. Also, set a good example. Children often mimic their parents, and if they see you return to life as normal, so may they. Your youngsters will pay close attention to how you're responding to events. That means you must deal with any emotional issues you may have following a terrorist incident. You might share your feelings with your children and tell them how you coped.

If problems persist, seek professional help for your child.

The danger of a traumatic event leading to severe emotional and academic difficulties is very real. Problems such as post-traumatic stress disorder (PTSD) or substance abuse could follow. If a child remains unable to cope with a traumatic event long after its occurrence, seek professional help. Consult with your child's doctor or school for recommendations, or contact the National Center for PTSD at www.ncptsd.org or 802-296-6300.

TIPS FOR THE ELDERLY AND DISABLED

Do not hesitate, therefore, to contact friends, relatives, and even neighbors you don't know about providing you with assistance in the event of a terrorist incident or other emergency. If some people reject your appeal for assistance, don't let it upset you. They'll eventually get their reward for their selfishness.

Fill out an emergency health-information card and keep copies around the house.

Communication with emergency personnel could be difficult following a terrorist attack or other crisis. No one can count on being conscious or coherent following a terror incident. If you are elderly or disabled and require special care, this could prove to be a serious problem.

The American Red Cross suggests that the disabled and elderly fill out an emergency health-information card. The card tells rescuers what they need to know about you if they find you unconscious or incoherent, or if they need to help evacuate you in a hurry. The card should list the medications you take and the health-care equipment you use, as well as any allergies or sensitivities you may have. Also list the names of people who should be contacted in an emergency. Make multiple copies of the card to keep in your purse or wallet, near the doors to your home, in your car, and if appropriate, in your wheelchair pack.

Keep a working flashlight handy in case you have to signal emergency personnel.

If you're disabled or elderly, you should augment your emergency supplies.

Make sure your have enough prescription medicines and other necessary medical supplies (e.g., medication syringes, colostomy, respiratory, catheter, padding, distilled water, etc.) to last one to two weeks in the event of an emergency. If you have a respiratory, cardiac, or multiple chemical-sensitivities condition, store towels, masks, industrial respirators, or other supplies you can use to filter your air supply. Be sure, too, that you have a nonelectric can opener and a battery-powered radio (and extra batteries) in case the electricity goes out. If you use a hearing aid, keep spare batteries on hand. And stay in contact by phone with your friends and relatives.

Follow the instructions of emergency personnel if you're asked to leave your home. Leave your valuables in a safe place at home rather than taking them with you. The chances are higher that you'd lose them as opposed to having them stolen from your unoccupied home. See that your pets are taken care of by a friend or neighbor, brought with you, or left at home with plenty of food and running water.

The American Red Cross says that if you are unable to afford extra supplies, you should consider contacting one of the many disability-specific organizations, such as the Multiple Sclerosis Society, Arthritis Foundation, and United Cerebral Palsy Association. These organizations may be able to assist you in gathering extra low-cost or no-cost emergency supplies or medications.

Establish and maintain a personal support network.

You may already have a personal assistant, but an emergency could make transportation impossible, preventing your helper from reaching you. As a backup, establish a personal support network of friends and neighbors who could lend a hand in an emergency. If you don't know many people in your neighborhood, contact a local church, synagogue, or mosque and ask if they could provide names of people living nearby who'd be willing to lend you a hand in a crisis. But don't just file those numbers away. Dial them every once in a while, saying hello and telling your network that you're practicing just in case an emergency should happen.

 Ask in advance if any of your contacts would be willing to take care of your pets in the event of an evacuation.

Discuss things in advance with the members of your support network.

The American Red Cross recommends the following seven important items you should discuss with and give to the members of your personal support network:

(1) Make arrangements, prior to an emergency, for your support network to immediately check on you after a disaster and, if needed, offer assistance.

(2) Exchange important keys.

(3) Show where you keep emergency supplies.

(4) Share copies of your relevant emergency documents, evacuation plans, and emergency health-information card.

(5) Agree on and practice a communications system, but don't count on the telephones working.

(6) You and your personal support network should always notify each other when you are going out of town and when you will return.

(7) The relationship should be mutual. Learn about each other's needs and how to help each other in an emergency. You could be responsible for food supplies and preparation, organizing neighborhood watch meetings, interpreting, etc.

Ask your helpers to read some literature on how to assist you in an emergency.

The Internet sites of the American Red Cross (www.redcross.org) and the Federal Emergency Management Agency (www.fema.gov), among others, have considerable amounts of literature on how to assist the elderly, the immobile, and the disabled in emergencies, including ones related to terrorism. Ask the members of your support group to familiarize themselves with the available literature so they can better prepare for an emergency and know instinctively what to do should a crisis strike.

✔ If you use a wheelchair or other special equipment, show your helpers how it operates. If you have difficulty speaking, ask someone to record a message you can use over the phone in an emergency; the message should include your name, address, and the nature of your disability.

Get your local police to conduct a home security check.

Police departments are well aware of the special needs of the elderly and disabled, and they're also aware that criminals often see the elderly and disabled as easy marks. Therefore, contact your local police department and ask an officer to conduct a security check of your home. This would have the added benefit of informing the police of your location and needs in the event of a local emergency.

+ CHAPTER 2 +

Survival and Escape Strategies

AMBUSHES AND ASSASSINATIONS

Senior executives, like top government officials and celebrities, often have bodyguards, professional drivers trained in evasive maneuvers, and even bullet- and bombproof cars. Not everyone can afford such protection, however. Given that terrorists sometimes attack midlevel executives or government personnel, it's worthwhile knowing how to spot and avoid ambushes and planned assassinations. Of course, if you think your life is in danger, contact the police or FBI.

Know the five primary means of assassination.

Terrorists typically murder their intended targets using bombs or gunfire in different ways. (1) A roadside bomb might be affixed to a utility pole, parked bicycle or car, construction barrier, detour sign, or pushcart. (2) A car bomb might be planted in the victim's vehicle. (3) Gunfire may erupt from a stationary location like a fake construction site or parked car as the victim passes by, or the victim may be gunned down as he exits or enters his car, home, or office. (4) Gunmen in a moving car or on a motorcycle may overtake a victim's car while he is driving. And (5) a terrorist may be working undercover as, say, a taxi driver and pick up his unwitting victim as a fare.

Look out for terrorist surveillance and reconnaissance.

Prior to an attack, terrorists will conduct surveillance of an intended victim to assess his habits, patterns, and routines, most especially his travel routes near his home and office. A reconnaissance team will then survey those routes, looking for choke points where a road narrows and vehicles are forced to move slowly. Be aware of any unusual activity that seems to be centered around you and the places you live and work.

Conduct your own neighborhood reconnaissance.

To better spot an impending attack, you need to develop an instinctive sense of your surroundings, particularly near your home and office. Take a drive to get an impression of what looks normal—so you can tell what isn't normal, should the occasion arise. Also, look for natural choke points that terrorists might exploit—narrow streets, a blind spot in a road, roadside construction, etc.—and then avoid those areas.

Recognize the curious signals of an impending on-the-road attack.

Here are a few of the ploys that terrorists sometimes use to abduct their target: striking your vehicle from behind with another car; cars (or even pedestrians) boxing you in at a light or stop sign; a fallen cyclist or pedestrian; a tree lying in the road; an overturned baby carriage or large barrier; phony flagmen, road construction, and detours; and persons dressed in fake police or security uniforms.

Get someone to watch your back if you think you're under surveillance.

If you feel the risk of attack or abduction is high, have someone watch your back to determine whether you're being tailed and targeted for an assault. Preferably, employ a professional to do this work. Otherwise, ask someone with enough brains and brawn to do the job without endangering himself to any substantial degree. With sufficient evidence of a plot against you, you should be able to get the assistance of law enforcement. Either way, if you are abducted while someone is watching your back, the police would likely have enough information to track down your abductors and discover your whereabouts, hopefully, in relatively short order.

On the road, don't stop to help strangers.

Terrorists (as well as carjackers) are known to use ploys like fake accidents and seemingly broken-down vehicles to get at their prey. If you come across a road accident or stranded motorist, don't stop to help. Instead, call for assistance using your cell phone or the nearest public telephone.

Take precautions by varying your everyday routines and routes.

Terrorists rely on a target's predictability and unvaried routines to carry off a successful ambush or assassination. You need, therefore, to mix up your transportation patterns and your daily schedule continually. Avoid taking the same routes every day, especially the streets and roads near your home and workplace. (The vast majority of terrorist ambushes and murders occur in proximity to the victim's home or office.) Also avoid narrow streets or places where construction is taking place and other choke points. Try to pick roads with multiple lanes and routes that allow for maximum speed. Don't drive near the curb; keep as close to the center of the road as possible. Vary your patterns when it comes to taking lunchtime strolls, picking up kids at school or day care, jogging, cycling, or going to the gym. Leave for work and return home at different times. Enter and leave work using various doors. Shop at different stores, and eat at different restaurants. Steer clear of unlit areas and, whenever possible, avoid bridges and tunnels.

> ✔ If there's a personal nameplate or Executives Only sign at your parking space at work, have it removed. Park in a different space every day.

Don't telegraph your exit from a building.

If you have a driver, don't let him idle at the door, waiting for you to leave a building. Have him pull up just as you are ready to leave (or employ a decoy car). Both the failed assassination attempt on President Ronald Reagan in

1981 and the successful bid that killed Israeli prime minister Yitzhak Rabin in 1995 occurred as the men were about to enter limousines that had been waiting at the curb for so long that crowds had gathered. This gave their attackers plenty of time to get into place and prepare for the attacks. By blending into the crowds of onlookers, the assassins caught security off guard. If, however, the attackers hadn't been tipped off by the sight of the waiting cars, perhaps they might have been caught before a shot was fired.

Know how to escape an ambush.

When driving, always leave enough room between you and the vehicle in front of you—most especially, whenever you come to a stop—to allow you room to maneuver. Give yourself enough room to turn your car around and head in the opposite direction if necessary. Otherwise, be willing to drive up on curbs, but try to avoid driving on slippery surfaces like grass.

✔ To ensure you have sufficient room to turn around, never get so close to a vehicle in front of you that you can't see its back tires touching the road.

Counterattack if you're under fire and escape is impossible.

The best way to counterattack a moving vehicle that is firing on you is by forcing it off the road. Turn your vehicle into the terrorist's car, then accelerate as you push it off the road or into a stationary object like a parked car, tree, or light stanchion. If your attacker is in a stationary car or behind, say, a construction barrier, ram it with your car, but only if escape is impossible. A motorcycle gunman (usually seated behind the motorcycle driver) racing by your vehicle would be seriously injured if not killed were you to open your car door as he passed. (Motorcycle assassinations became so rife in Italy and Colombia that both nations have banned second persons from riding on the backs of motorcycles and scooters.)

Never take the first taxicab in line.

Terrorists have been known to take jobs as taxi drivers—both to give them cover as a sleeper in their target country and also to carry out assassinations and other assaults. It's wise, therefore, never to take the first cab in line when leaving a restaurant, hotel, or office, especially if you think you may be have been targeted by a terrorist group. Hail a moving cab instead.

Know how to inspect your vehicle for car bombs.

If you feel you may be targeted by a bomber, inspect your vehicle routinely. A proper inspection by one person normally takes 10 to 15 minutes. First, you need to become familiar with every part of your vehicle, including the engine compartment and undercarriage of your vehicle, so you'll notice anything that may be out of the ordinary. Get a sense of what looks normal. Use a flashlight and a mirror to aid in the inspection. Next, know what you're looking for in bomb detection. Telltale signs of bombs include strange wiring, cut wires, tape, lengths of fishing line, puttylike lumps (plastic explosive), boxes or other containers (concealing, perhaps, sticks of TNT), shunts from detonators, and unexplained smudge marks or grease.

Check the area around your parked vehicle, widening your search to a distance of 30 to 40 feet. Survey around and beneath the vehicle for discarded bits of tape, wire, or wire insulation. Look for grease, scratches, or scuff marks on the vehicle's exterior. Inspect for wires, foil, or objects attached to the outside of your vehicle. Pay attention to the grille, hood release, bumpers, tires, hubcaps, wheel wells, gas cap, exhaust pipe, door seams, and door locks. Check locks, windows, and weather stripping for forced entry. Look under the car (preferably using a mirror), inspecting the chassis, exhaust pipes, drive train, gas tank, bumpers, and areas beneath the trunk and seating compartment, especially for foreign objects.

Before opening the doors, peer inside the car for any suspicious items or signs of intrusion. Note whether the angle or position of the seats has changed since you were last in the car. Look to see if anything is behind the rearview mirror or if the mirror's angle has changed. Check to see whether floor mats or items left on seats have been disturbed.

Carefully examine the seams around doors for wires before opening them.

From outside the vehicle, check under the seats, headrests, and dashboard (especially the ignition wiring). Check the glove compartment and ashtray; gingerly look behind the sun visors and in any door pockets. Pop the hood and inspect the engine compartment, paying close attention to the wiring and the fire wall. Look for new wires, cut wires, tape, new-looking objects (e.g., oil filter), foreign materials, and unexplained grease marks, scratches, or smudges. Finally, inspect the seam around the trunk for wires; then, look inside.

 It's wise to leave glove compartments and ashtrays open after parking your car. Also, leave the sun visors down. Keep an item on the front seat, such as a newspaper or tissues box, so you can see if it has been moved since you parked the vehicle. Install a lockable gas cap, car alarm, and protective screen for your tailpipe. Whenever possible, keep your vehicle in a locked and alarmed garage.

Don't put pressure on seats with your hands or knees when inspecting underneath for an explosive device; some detonators are pressure-sensitive.

DEALING WITH VIOLENCE

Most people aren't prepared to deal with random acts of violence—be they criminal or terrorist in nature. Thinking the problem through and knowing how to respond beforehand can greatly lessen your chances of serious injury or death in a violent confrontation.

Never make eye contact with an assailant.

Whatever you do, don't ever make direct eye contact with an assailant or anyone who you suspect might do you harm. Eye contact raises an attacker's stakes in a confrontation, and he's likely to become more aggressive as a result. Not only is eye contact intimidating to an attacker because it represents a personal challenge, it also reminds him that he could be identified if he's later caught by the police. He's thus likely to become more violent.

Choose flight over fight.

An assailant represents an unknown entity. You have no idea what his true intentions are, how powerful he is, or how determined he is to do you harm. Our natural instinct when attacked is either to flee or fight; the prudent course is to flee. The problem is that our pride can sometimes cloud our judgment. Unless you're properly trained in the art of self-defense, it isn't sensible to try to fight off your attacker if fleeing him would be just as easy. Flight, of course, isn't always an option. You may find yourself locked in a confined space, such as an elevator, or you might be surrounded by a group of thugs. Still, if there's any chance of escape, take it.

Scream if someone tries to mug you or abduct you.

If you're attacked, it's better to attract attention than to be bundled off quietly. Sometimes, people hesitate to yell because they think it will upset their attacker. Well, that's precisely what you want to do! You want to throw him off stride. You want to give him something to think about. Your attacker, more than likely, has rehearsed the attack in his mind. He thinks he has everything figured out. He knows precisely what he intends to do. But, then, you disrupt his plans by screaming. For a moment, he's not sure how to respond. And at that very moment you have the advantage and a chance of escape. Your screams have thrown your attacker for a loop. So, by all means, scream. And get over any idea that screaming may upset your attacker; he is planning to harm you.

Scream "Fire!" You'll get more attention that way.

The best way to get help when you're being attacked is to gain the attention of people nearby. It's a funny thing about human nature, though. If you cry out for someone to call the police, people most often pull back, crablike, into their shells. They give your cry for help a second thought. "Does she really need help?" they wonder. "Is this just a girlfriend-boyfriend tiff? A wife-husband dispute? Do I really want to get involved?" Odds are they won't do anything, leaving it for someone else to call the police. So don't yell, "Help! Police! Police. Somebody, call the police!" Instead, scream, "Fire! Fire! Help! There's a fire! Fire!" Such a request elicits a totally different response from people within the sound of your voice. They will most likely call the fire department—as opposed to the police—and then they'll rush to see what's the matter.

Why do people do this? Why do they respond more positively to screams of "Fire!" than to pleas for police help? It's boils down to self-interest. An attack on you isn't a direct threat to the others around you. And when you ask them to call the police, it means that you're asking them to get involved in something in which they aren't an immediate party. But when fire is involved, the whole situation changes. Fire is a threat to *them,* a threat to the people living and working in the surrounding neighborhood. That makes their response to your screams different. They *will* call the fire department, because they care about *themselves.* By screaming "Fire!" you turn their self-interest to your personal advantage. It's not your welfare they care about so much; it's their own safety and property that they're worried about. They don't want to see their homes and businesses go up in flames; they don't want to be trapped in a burning building. So they'll call their local emergency number to protect not you but themselves.

A scream of "Fire!" does one other thing. It's likely to catch your attacker completely off guard. As many times as he may have rehearsed his crime in his mind, as often as he has attacked people before, a scream of "Fire!" is totally new and unexpected. It's something out of the blue, out of the ordinary. Indeed, if your scream is really effective, your attacker may even stop to look around to see where the fire is. Now, that's just what you want. You want that split second to flee your attacker—and have a good laugh to boot.

Consider taking a self-defense course to boost your confidence.

While we don't necessarily recommend that you try to overpower an attacker (unless you are especially gifted in martial arts), just the knowledge that you can defend yourself can be a significant morale booster. More important, if you expect to make trips to parts of the world known for their high crime rates, some basic training in self-defense is a must. Never forget, however, that a defensive action on your part will elicit a response from your attacker. Usually, his reaction will fall into one of two categories: fight or flight. If an attacker himself comes under assault, he must decide whether to fight back or run away. It's impossible to predict which choice he'll make with any certainty. On the one hand, if he has a violent temperament, an aggressive action on your part might only make the situation worse and put you further at risk of serious bodily harm. On the other hand, if your attacker is a scared kid, he may opt to run away if you display any talent in self-defense. Amateur assailants are often highly nervous and frightened; hardened criminals usually aren't.

Don't get us wrong. By all means, resist an attacker if he means to do you physical harm or attempts to abduct you. Put up a struggle. And yell for help. But getting into a fight with an attacker might not be the best choice in every circumstance. It all depends on the nature of the attack, the capabilities of your attacker, and how good you are at self-defense.

SURVIVING AS A HOSTAGE

It may have been planned that way or it may be the result of a botched attack, but too often innocent people find that they've suddenly become hostages to terror. If you find yourself a hostage, it's important to remain calm. Don't do or say anything—especially in the first five minutes when an abductor is most on edge, but know that there are things you can do to improve your chances of survival.

Never look your assailant in the eye.

Terrorists (and criminal kidnappers) are extremely nervous, especially in the early minutes of their action. Nerves sometimes give way to rash decisions.

A terrorist is most likely to kill a hostage early on in an attack—usually is to make a point to the other hostages that he means business and to show authorities that he's willing to do anything. A human life is of little consequence to him. It's especially important, therefore, that you never look your abductor in the eye. Direct eye contact is often viewed as attempt by a hostage at intimidation or defiance. The defiant ones run a high risk of being killed.

Don't do anything that would make you stand out from your fellow hostages.

Terrorists often feel a need to demonstrate their power and control over their hostages and send a message to the authorities. This is another reason they may decide to make an example of one of their captives by murdering him in cold blood. The person may be picked at random. Alternatively, the person may have invited retaliation for having insulted his captors in some way. Or he may simply have said or done something that made him stand out in the terrorists' eyes. So, keep a low profile if you're taken hostage. Go along with whatever the terrorists may want you to do. Act passively and be cooperative.

 Don't ever try to negotiate with terrorists or attempt to cut some kind of a deal.

If your odds of surviving the hostage crisis seem low and you see a chance to escape, take it.

Most times it's wise just to stay put as a hostage. The authorities typically have the site surrounded. Surveillance devices are probably being used to listen to and perhaps watch the terrorists. Sharpshooters may have the terrorists in their sights. However, in some incidences your chances of survival are low. That's especially true if the terrorists are bent on suicide, and the only reason they haven't already killed their hostages and themselves is that they are playing to the media, airing their demands and hoping to stir up more fear and anxiety. In such a case, if you see an opportunity to flee, take it. It may be your only chance at survival. Think your escape route through, and time your move. Flee before the terrorists get a chance to tie you up or you become exhausted by a long ordeal. Wait for the hostage takers to become distracted. Move quickly and silently. And don't look back. If you hear a demand from the terrorists for you to stop, freeze. Don't move a muscle until you're told to. And don't say anything other than "Okay."

For more information on hostage and kidnapping survival, see chapters 5 and 6.

GUNFIRE, GRENADES, AND BOMBS

Most civilians don't know how to respond properly to the sounds of gunfire, hand grenades, or bombs. Fight or flight is our instinctive choice. In cases of gunfire and explosives, however, neither of these responses may save your life. You have to discipline yourself, perhaps through role-playing or the use of your imagination, to think before reacting in such situations.

Don't automatically run from the sounds of gunfire; think first of hitting the ground.

The best response in many situations involving gunfire (and explosives) is simply to drop to the ground, lying facedown. Don't necessarily run, because that could expose you to stray bullets. Gunmen, moreover, tend to look for targets in their line of sight—and that includes fleeing pedestrians or building occupants. By your hitting the deck, a gunman may overlook you or think you're already dead. Bombs, too, do the most damage to people nearby who are upright.

✔ Stay put until instructed by authorities to move. Otherwise, you risk being caught in cross fire or possibly being mistaken for a terrorist.

Assume a tucked position to protect your body.

Protect your vital organs and major arteries by pulling your upper arms and elbows into your sides, thus guarding your heart and lungs, and cupping the palms of your hands over your ears, thereby covering the arteries in your neck and protecting your head and hearing.

Drop to the ground, too, at the sight of a hand grenade or bomb.

Attempting to flee a hand grenade or bomb left on the street or carried by a suicide bomber could cost you your life, because grenade and bomb shrapnel flies outward and upward in a cone shape. Hitting the ground immediately could save you from serious injury or death.

Know the exceptions to the rules involving street terrorism.

There are exceptions to the general rule of diving to the ground in an act of street terrorism. If, for example, you're only a step or two away from a protective barrier, such as a vehicle, solid pillar, heavy planter, or the corner of a wall or building, you ought to get behind it as quickly as possible. If you have children with you, grab them and either drag them behind a barrier or drop on them, covering them with your body. Finally, if the terrorist is screaming demands or shouting slogans, this may give you time to flee before he acts.

As noted earlier, a facial disguise is tip-off to a terrorist's intention. Wearing a disguise usually means he plans to survive the attack and not blow himself up in a suicide bombing. What this tells you is that you probably have time to flee to safety.

Learn the proper ways to crawl or roll to safety.

The way most people crawl would expose them to bodily harm in a terrorist street incident involving gunfire or explosives. Their backs and behinds would likely be up in the air, exposing their spines to potentially lethal or paralyzing injury. There are three recommended techniques, employed by most militaries, of staying low and getting to safety under fire.

LOW OR BELLY CRAWL   Keep your body flush to the ground, turning your head sideways. Begin by pushing both arms forward and bending your right leg, pulling it forward until the knee is as far as it will go. Then, move forward by pulling your body with your arms and pushing against the ground with your right leg. Continue this push-pull movement until you reach safety. This technique may be slow, but it greatly reduces your body's exposure to bullets and shrapnel.

HIGH CRAWL   With your chest off the ground, rest your weight on your forearms and lower legs. Extend your knee well behind your buttocks, thus lowering the profile of your back and spine. Crawl forward by alternately advancing your right elbow and left knee, then your left elbow and right knee. This technique saves time but exposes you to some danger. It should, therefore, be used only in incidents where gunfire isn't coming in your direction.

ROLL   With your arms at your side or over your head, roll along the ground to safety. Obviously, this technique isn't intended for long distances, but it's valuable to know if you need to roll, say, under a nearby vehicle or piece of furniture.

 Stay away from glass or heavy objects that could topple over in a blast.

After an incident, beware of a possible second bomb in the area.

In Israel, Palestinian terrorists make it a practice to plant a second explosive device, usually in a parked car, in the vicinity of an initial attack. The aim is to kill or injure emergency personnel, as well as curious onlookers attracted to the initial explosion. Therefore, clear the area following a detonation. Don't let your curiosity get you killed.

Get as much information as possible if you receive a bomb threat.

If you get a call from someone saying a bomb (or a chemical, biological, or radiological device) is about to go off, stay calm and try to get as much information from the caller as possible. Keep the caller on the line as long as possible and get someone to call the police using a landline phone and not a cell phone. Indeed, have everyone turn off all cell phones and radios, which could trigger a bomb. Write down everything the caller says. Ask him where the device is located, what kind of device it is, what it looks like, and when it's supposed to go off. Make notes of the caller's vocal characteristics and apparent emotional state (e.g., calm, giggling, or distressed). Evacuate the premises. Don't touch any suspicious-looking packages. Inform the police that you took the call, because they'll want to debrief you. (See chapter 8 on business and building security for a detailed bomb-threat checklist.)

 While evacuating, again, steer clear of heavy objects that could fall over and glass that could shatter.

EXPLOSIONS, FIRES, AND COLLAPSES

The events of September 11 show the importance of knowing how to escape from a building that has been attacked by terrorists or surviving if you've been pinned in a collapsed structure.

In a building explosion, immediately get under cover.

Assuming you're not in close proximity to a bomb, your greatest dangers in a building explosion come from flying glass and falling objects. At the sound of an explosion, dive under the nearest desk, table, or chair—or at least hit the floor—to avoid the shrapnel-like effects of the blast. Be especially careful if you're near a heavy object that could fall over and crush you. Once it's all over, leave the building, then proceed a good distance away for fear that a second explosion might occur outside. Cover your mouth and nose with a handkerchief or cloth, wetting it down if possible, to reduce the inhalation of dust, smoke, and debris.

✔ If you're in a wheelchair, it's advised that you stay in it in an explosion. Cover your head with your hands, then ask someone for help in evacuating the premises.

In a department store or other crowded public place, be careful not to get trampled.

Should an explosion occur in a crowded public place like a department store or transportation hub, your biggest worry—apart from flying glass and falling objects—is stampede. Panic would likely follow an explosion (or, say, gunfire). This could result in many people being trampled or pinned against doors that won't open. Once the dust has settled, your best bet is to get out of the flow of traffic. Get behind a counter or next to a wall or pillar. Scan the area for exits that aren't blocked. Pick the quickest route, and exit the building.

Err on the side of caution in an emergency; it's better to be safe than sorry.

We've heard of one tragic case at the World Trade Center in which a young woman told her coworkers, "I'll be right with you," as they headed for the emergency stairs. The woman was never seen again. She said she wanted to make "one last call" to a client in London. That ill-fated decision cost the woman her life. Such a sad tale should serve as a reminder to everyone faced with an emergency to act and act quickly.

Don't dally. Nothing, absolutely nothing, is as precious as life. When an emergency evacuation is called for, get out of the building as quickly and as safely as possible. Apologies can always be made later for a delayed phone call or a late e-mail. Whatever you leave behind at your desk will most likely be there when you get back. At worst, you'll have to buy a new briefcase, cell phone, or laptop computer—but at least you'll be alive to go shopping.

Don't follow instructions that don't instinctively sound right.

Numerous survivors of the Twin Towers attack reported hearing repeated announcements over the public address system in one of the buildings, telling workers that the building was safe and to return to their offices. Some people followed that wrongheaded advice. One can only imagine their anguish (assuming they had time to reflect) once they realized the horrible mistake they'd made. Bad advice had sealed their fate. Others, many others, ignored the recommendation to return to their offices and instead continued down the staircases to safety below.

Although no one can be sure, odds are that most of the survivors simply decided to follow their instincts rather than the anonymous advice being blared over loudspeakers. It's a lesson worth remembering.

Each of us comes equipped with natural instincts that help to keep us alive and safe from harm. No one has to tell us, for instance, not to step off the edge of a cliff or into the path of a speeding car. We have an innate sense of what not to do. Only reason, or better, rationalization, can countermand our natural instinct for self-preservation. So, if it comes down to a split-second decision between following your head or your gut, go with your gut. Continue to do what you were doing in the first place. Do what your whole being is telling you to do. And don't second-guess yourself. Don't try to figure it all out. There isn't time. Just get out of harm's way—and think about it later.

Survive a fire by staying low, protecting your breathing, and moving quickly.

Try to evacuate the premises immediately in case of fire. Shout to alert others to the danger. Don't waste time trying to save property; it could cost you your life. Fire can travel at lightning speed—up to 19 feet per second. Before opening a door in a fire emergency, feel it first for heat. Place the palm of you hand on the door, around the cracks, and on the doorknob. If it's hot to the touch, don't open the door. Use another escape route. Close but don't lock doors behind you as you leave. Stay low to avoid any smoke. Smoke and poisonous gases first collect along the ceiling, so stay below the smoke level at all times. Knock on any doors you may pass to alert others to the fire. Leave by the nearest fire exit or stairway.

✔ Never use an elevator in a fire; the doors could open on a fire floor, incinerating you. And never go back into a burning building; the odds are you won't come out again alive.

Protect yourself from smoke and flames if you're trapped.

If you're trapped in a burning building, use tape or moistened cloth to seal off the airflow from around and beneath the door, thereby reducing the amount of smoke that penetrates your room or office. Smoke inhalation kills more often than a fire's flames. Go to the window and signal for help. Be careful about opening windows fully, for the draft could suck smoke into the room. Instead, open the windows just a few inches from the top or bottom to get fresh air. If a phone is available, call 911 and tell the operator your precise location in the building. Otherwise, dangle a sheet or piece of clothing from the window. Turn off the air-conditioning if you can. If water is on hand, wet down the door nearest the fire. You also can moisten cushions or mattresses and then prop them up against the fire door, using furniture to hold them in place and repeatedly dousing them with water.

For additional fire survival and safety information, see the Internet pages of the U.S. Fire Administration (www.usfa.fema.gov), Los Angeles Fire Department (www.lafd.org), and New York City Fire Department (www.nyc.gov).

If trapped in debris, conserve energy but still make your location known to rescuers.

Try to be calm if you're trapped in debris following a building collapse. Conserve your energy. Concentrate on your breathing to help calm your nerves and lower your blood pressure. Don't move around because that would accelerate and possibly impair your breathing. Use whatever means is at hand to signal your location to rescuers. Bells, flashlights, whistles, or personal alarms are great if available. Otherwise, tap regularly on a wall or pipe. Try not to shout too much, for it will weaken you and cause you to lose your voice and inhale large amounts of dust. Yell only as a last resort, especially if you hear the sounds of rescuers nearby. Urination could help rescue dogs pick up your scent.

 It's advisable that untrained persons not attempt to rescue anyone trapped in a damaged building, for the chances of a secondary collapse are high. Leave the job to emergency personnel.

Where there's time, there's hope.

Unless you are so unfortunate as to die instantly from a terrorist attack or other calamity, you normally have some time, however brief, to act. And that could make all the difference in terms of your personal survival. Never give up hope or the will to live.

BIOHAZARD AVOIDANCE

Biological weapons are so stealthy that they could show up in many guises. These include food and perhaps even fragrance samples. No advance warning of such an attack will be given. It could indeed take days or weeks before anyone realizes what has happened. Therefore, you need to take prophylactic steps to avoid exposure to biohazards.

Avoid seemingly accidental liquid spills or anyone using a spraying device.

Terrorists could spread toxic chemicals or communicable diseases via mists from handheld spray bottles, gas canisters, or even overturned 50-gallon drums. Over large areas, a truck-mounted insect sprayer would be effective yet difficult to detect as being anything other than a routine activity. If you spot a team of suspicious-looking cleaners, say, hauling a large drum or spraying equipment in a crowded public building, move away and alert the police or building security.

Be wary of any spraying activity that takes place at night and hasn't been preannounced by local officials.

Throw away free samples of fragrances or food.

Terrorists are ruthless by nature. The usual rules of fair play don't apply as far as they're concerned. They can and will do anything that they think will kill, injure, or otherwise frighten their targets. Chemical and perhaps biological warfare agents are in their arsenal of weapons (as well as the arsenals of some of the nation-states that support them). The big question for terrorists then becomes how to vector those agents to their targets.

A vector is simply the name given to the means by which a chemical or biological agent gets to its intended target. Conceivably, even a person who was wittingly or unwittingly exposed to an infectious disease could be used as a vector of biological terrorism. Poisoning samples of free food, say, delivered in the mail, would be one way.

Another perhaps less obvious but more insidious way would be to disguise a biological agent as a perfume or a cologne sample. Take the scratch-and-sniff strips or the pull-open strips for fragrances commonly found in magazines or promotional mailings. If terrorists had a biological agent that, for example, activated only after it came in contact with oxygen, a fragrance strip would be a perfect vector. It wouldn't take much doing for terrorists to set up a dummy company, ostensibly selling food products or fragrances, and use it as a means of distributing poisons to the general public.

The best means of preventing such an attack would be to deny terrorists the potential vectors of food and fragrance samples by banning these items from the postal system, magazines, and newspapers. You can only imagine the horror and carnage that would be caused if people throughout a city or county starting getting ill, collapsing and dying after smelling a perfume insert that had come in their Sunday newspaper. The risks are simply too great for the government (and governments globally) not to take preventative action against such a stealthy terrorist threat and for publishers not to voluntarily do the same. Until then, the adage "Let the buyer beware" takes on a new significance.

Forgo eating food from salad bars or restaurant smorgasbords.

Terrorist may decide to target eateries in large cities, financial districts, or airports, or food establishments near government facilities and key businesses. A relatively easy way to poison food would be to lace open salad bars or unattended smorgasbords with biological (or chemical) agents. It's best, therefore, to avoid eating any food that has been within reach of the public.

Exercise caution when buying food from street vendors or roadside stands.

Terrorist "sleepers," who might be resident in a country for years and not show any signs of being anything other than hardworking immigrants, may have taken positions in the food industry as part of a plan to disseminate biological agents. That's not to say that every foreigner working around food is a potential terrorist. But it does mean that you should exercise caution in purchasing food from establishments where there is little or no supervision, such as sidewalk vendors and roadside stands.

Be aware of common food-borne illnesses.

In 1984, a religious cult in Oregon contaminated restaurant salad bars and water glasses with salmonella bacteria, hoping to affect the results of a local election by incapacitating voters. No one died, but 751 persons fell ill with food poisoning. The U.S. government has since become so concerned that terrorists may contaminate the food supply that in early 2002 the Food and Drug Administration (FDA) urged the nation's food industry—from farmers and fishermen to food importers and restaurateurs—to beef up security. The recommendations included scrutinizing visitors more closely, performing background checks on potential employees, and inspecting incoming and outgoing vehicles for suspicious activity.

Common food-borne illnesses are caused by *E. coli* and *Salmonella* bacteria, among others. Symptoms include acute intestinal distress with the sudden onset of headache, fever, abdominal pain, diarrhea, nausea, and sometimes vomiting. In severe cases, the victims can die. Young children and the elderly are at particular risk. If you believe you've eaten tainted food, contact a physician or emergency room immediately. You can also call the U.S. Department of Agriculture's special meat and poultry hot line at 800-535-4555. For incidents involving food other than meat and poultry, call the FDA's 24-hour emergency number, 301-443-1240.

Examine store-bought food carefully for signs of tampering.

Food that has been tampered with will often change in appearance and texture. It may discolor or degrade quickly; it may become slimy to the touch and develop a foul smell. Cans containing contaminated food may bulge or buckle, and the metal tops of jars or bottles also may bulge. Beware of broken seals and punctured or undone plastic wrapping around fresh meat and poultry.

Take precautionary steps in the kitchen.

Wash all fresh foods thoroughly, including meat, poultry, and produce. Remove the outer layers of fresh vegetables and fruit. For those most concerned about bioterrorism, wash produce in soap and water or rinse in a highly diluted chlorine-bleach solution. Cook foods thoroughly.

Wash your hands after returning from an outing, most especially if you used mass transit or a taxicab.

If a biological attack occurs and the symptoms take time to develop, the disease could spread far and wide in a variety of ways. Exchanging handshakes can transfer germs. The common cold often gets transmitted from person to person via the hands. But there are other, less obvious routes of disease transmission. Philip M. Tierno Jr., author of *The Secret Life of Germs: Observations and Lessons from a Microbe Hunter,* notes that "buses, trains, and taxis abound with surface, such as handrails, that act as collecting areas and transfer points for germs." He advises that during and after traveling on mass transit or in taxis, you not touch your face, eyes, nose, or mouth until you've had a chance to wash your hands. He even urges people to use tissues or paper towels when opening and closing doors in public buildings, especially washroom doors.

✔ Tierno offers another interesting piece of health advice: wear long pants or skirts when visiting public facilities, such as movie theaters, sports arenas, or concert halls, because sitting down with bare legs could lead to the contraction of an infectious illness from germs on the seats.

Stay away from persons who are purposely coming in close contact with strangers.

Given the fanatical and even suicidal nature of today's terrorists, nothing can be ruled out, including what can only be called walking bombs. This doesn't mean only suicide bombers with explosives strapped to their bodies. A terrorist could expose himself (or be exposed unwittingly) to a communicable disease, such as smallpox or plague, and then mingle with hordes of people to spread the disease. Beware of anyone in public who seems interested in getting close to strangers.

✔ Be careful, too, whenever someone approaches you asking for information or if an argument or street scuffle breaks out near you. Pickpockets often employ helpers to distract their targets.

Stock up on inexpensive face masks in case of a biochemical attack.

Experts at the Oak Ridge National Laboratory have found that commonly used N95 face masks (costing under $2 apiece) are effective, as are HEPA (high-efficiency particulate air) masks, against the inhalation of chemical and some biological warfare agents, such as anthrax. N95 masks block 95 percent of particles that are three microns or more in diameter. Keep the masks at home and at work, as well as in your briefcases, purses, and carry bags, and make sure that all members of your family have them.

If you buy a gas mask, get the right one.

U.S. authorities and others tried to dissuade the public from buying gas masks following September 11 and the subsequent anthrax outbreaks. On the other hand, most every Israeli has a gas mask (and knows how to use it), because of the continuous danger of an attack using chemical or biological weapons. Buy a gas mask if it makes you happy. It may not do you much good, however. First, you'd have to carry it with you at all times. Terrorists won't be giving advance warning of a biochemical attack. Second, even if you have one on hand, you could find that you'd donned it too late to prevent exposure to the toxic agent. Finally, protection against many types of biochemical weapons requires a complete outfit, covering the entire body from head to foot. What's more, any small gap or opening could let ambient air in and mean death or injury.

If you do buy a gas mask, get one that snugly fits around the contours of your face, neck, and head. Otherwise, it's a waste of money. And don't buy old army surplus; gas masks and filters are only good for so long. A lot of the equipment being sold since September 11 is useless. Buy masks and filters only from reputable dealers; some offer masks made in Israel, which tend to be quite good. Note, too, that masks are made for children and infants, as well as adults. For a detailed discussion of protective gear, see "How Do I Know? A Guide to the Selection of Personal Protective Equipment for Use in Responding to a Release of Chemical Warfare Agents," published by the Oak Ridge National Laboratory and available at http://emc.ornl.gov/emc/PublicationsMenu.html.

Weigh any decision to be vaccinated against a biological agent, such as anthrax, carefully.

Anthrax vaccine was made available to postal workers and others exposed to the contaminated letters that were sent anonymously in the U.S. mail in the fall of 2001. This followed a 60-day regimen of treatment with antibiotics. It's as yet unclear how effective such vaccines are and what the side effects may be. Before making a decision, discuss the matter with your doctor.

Pregnant women have been advised not to take the anthrax vaccine, because it can apparently cause birth defects.

NUCLEAR REACTOR EMERGENCIES

Radioactive materials would be life-endangering if terrorists successfully attacked a nuclear power plant or similarly struck a depot containing spent fuel rods or radioactive waste.

Become familiar with the terminology used to describe the different types of nuclear reactor emergencies.

Commercial nuclear power plants have a system for notifying the public if a problem arises. There are four emergency classification levels:

(1) *Notification of Unusual Event* is the least serious. It only means that emergency officials have been notified of an event at a nuclear power plant but the incident poses no threat to the public or plant employees. Therefore, no action on your part is necessary.

(2) *Alert* indicates an incident has occurred that could reduce a nuclear plant's safety level, but backup systems are still working. Emergency agencies are notified, but no action by the public is deemed necessary.

(3) *Site Area Emergency* means that major problems with a plant's safety systems have progressed to the point that a release of some radioactivity into the air or water is possible. However, the release isn't expected to exceed federal Protective Action Guidelines beyond the site boundary. Thus, no action by the public is said to be necessary.

(4) *General Emergency*, the most serious of the four classifications, means a nuclear plant's safety systems have failed and radiation could be released that would travel beyond the site boundary. State and local authorities would take action to protect residents living near the plant. People in the affected areas might be advised to evacuate promptly or

shelter in place. When the sirens are sounded, you should listen to your radio, television, or tone-alert radio for information and instructions.

If a warning sounds, don't panic.

An alert doesn't automatically mean evacuation. Tune to your local television and radio emergency stations for information and instructions. Note, for instance, that the sirens could be warning of a tornado, fire, flood, chemical spill, or other local emergency. Authorities advise against dialing 911; other emergency numbers will be provided.

Don't race to pick up your kids at school; call first, because they may already have been evacuated.

If an incident involving an actual or potential radiological release occurs, authorities plan to put the safety of schoolchildren first. If an emergency was declared, students within a 10-mile radius of the incident site would be relocated to designated mass-care facilities in a safe area. Usually, as a precautionary measure, schoolchildren are relocated prior to the evacuation of the general public.

Follow your evacuation plan if instructed to leave the area.

If an evacuation is mandated, put your emergency-supplies kit of water, food, and other provisions in your vehicle, also taking along your cell phone, credit cards, cash, prescription medicines, extra eyewear, and games for your kids. Follow officially designated evacuation routes. If the roads are jammed, fall back on your alternative exit plan. Stay calm, and don't speed. Consider giving a neighbor a ride out of the area. Be especially mindful of any children who may be home alone, the elderly, and the disabled. If you don't have a car, public transportation is expected to be available, so go to your nearest bus stop or train station.

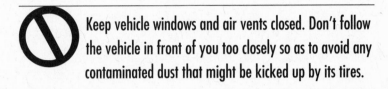

✔ Before leaving home, lock all windows and doors; turn off the air-conditioning, forced-air heating or cooling, fans, and furnace; and close fireplace dampers and any outdoor vents. These steps will help to minimize radiation contamination.

🚫 Keep vehicle windows and air vents closed. Don't follow the vehicle in front of you too closely so as to avoid any contaminated dust that might be kicked up by its tires.

Minimize your risk by putting time, distance, and shielding between you and the site of a radiation incident.

"There are three factors that minimize radiation exposure to your body: time, distance, and shielding," advises the Federal Emergency Management Agency (FEMA). Here are its explanations and recommendations:

TIME   Most radioactivity loses its strength fairly quickly. Limiting the time spent near the source of radiation reduces the amount of radiation exposure you will receive. Following an accident, local authorities will monitor any release of radiation and determine the level of protective actions needed and when the threat has passed.

DISTANCE   The more distance between you and the source of the radiation, the less radiation you'll receive. In the most serious nuclear power plant accident, local officials will likely call for an evacuation, thereby increasing the distance between you and the radiation.

SHIELDING   Putting heavy, dense materials between you and the source of the radiation will provide shielding and reduce exposure. This is why local officials could advise you to remain indoors if an accident occurs. In some cases, the walls in your home or workplace would be sufficient shielding to protect you for a short time.

✔ Travel upwind of the source of radiation. Also try to put hills, mountains, forests, or groups of tall buildings between you and the incident site.

If advised to stay indoors, take refuge in your safe-haven room or basement.

Begin by bringing children and pets inside. Anyone who has been outdoors should shower thoroughly, paying particular attention to hairy areas of the body. Remove all clothing and shoes worn outdoors and store them in sealed plastic bags; if contaminated, they will need to be destroyed. Pets that have been outdoors should be isolated in a room apart from people, with ample food and running water; remove any hazardous substances from the room. Consider if any neighbors may need assistance. Lock all windows and doors; turn off air-conditioning, forced-air units, fans, and furnaces; close any outside vents, including fireplace dampers. Bring a phone, battery-powered radio, candles, matches, and all other emergency supplies into your safe-haven room or basement. Seal windows and doors with tape and plastic sheeting.

🚫 Following a radiation incident, even one that occurred a good distance from your home, avoid eating food grown in your garden. Contamination can affect areas many miles from the accident or terrorist attack site.

Make an informed decision on potassium iodide.

Potassium iodide is a salt, similar to table salt. In fact, potassium iodide is used to iodize table salt. Its chemical symbol is KI. The U.S. Nuclear Regulatory Commission (NRC) requires states where people live within 10 miles of a nuclear power plant to consider stockpiling potassium iodide as "a protective measure for the general public in the unlikely event of a severe accident." The distribution of potassium iodide would supplement the usual protective measures of shelter and evacuation in a nuclear reactor emergency.

MASS DESTRUCTION ATTACKS

From the perspective of survival, you needn't separate the four classes of mass-destruction weapons—i.e., chemical, biological, radiological, and nuclear—into individual categories. Two will do. Although each of these four weapon types is unique, three of them kill and injure in similar ways. Chemical, biological, and radiological (CBR) weapons are similar in many respects when it comes to survival tactics. Nuclear bombs, however, are different. They emit energy in the form of blast, thermal radiation, and nuclear radiation and thus require special survival techniques.

This observation is important for it means that you need differentiate between only two broad categories of terrorist weaponry in order to know how to respond in the event of a mass-destruction attack. This foreknowledge could save vital time. It means, for instance, that you needn't identify the exact agent used to know what actions to take. Clearly, the brilliant flash of light, violent wind, and ominous sound of a nuclear blast aren't to be confused with a chemical, biological, or radiological weapon (even one dispersed using a conventional explosive). We'll, therefore, deal with CBR weapons and nuclear bombs separately in instructing you on how to survive. Let's begin with CBR agents. (Take note that the survival techniques for a CBR attack also apply to a terrorist incident at, say, a hazardous-chemicals facility that might hurl contaminants into the air and coat a neighboring area.)

First and foremost, protect your breathing in a CBR attack.

CBR weapons kill and injure most often as result of the inhalation of the toxic agent, so the first and most important step to take is to protect your breathing. Cover your mouth and nose with either a mask or a cloth (e.g., a handkerchief, towel, or piece of clothing) as you evacuate the contaminated area.

Cover up by donning layers of protective clothing.

Cover all exposed skin surfaces as much as possible by putting on layers of clothing, including a thick overcoat, sturdy boots, heavy gloves, and a hat. But don't waste too much time gathering up clothes, especially if the incident occurred near your location.

If you're outdoors and some distance from the attack site, head upwind.

Again, the aim is reduce your exposure to any toxic agent you might inhale, ingest, or get on your skin. You, therefore, want to get out of the way of any airborne agent being disseminated by the wind. So, travel with the wind blowing in your face. Close your vehicle's windows, and seal air vents. Don't tailgate, because you don't want churned-up dust getting into your vehicle.

✓ If you can't tell which way the wind is blowing, do what golfers do: drop something light from your fingertips and watch which way it drifts. Don't, however, pull up blades of grass, as golfers are wont to do, for the grass could be contaminated. Instead, use a bit of paper or tissue that may be in your purse or pocket. Even lint or hair will do.

If you're indoors and some distance from the attack site, stay put.

Since CBR agents can be taken by the wind, being outdoors following a mass attack always entails some risk. If you're a good distance from the attack site, say several miles away, your best bet is to remain indoors. Roads will be chaotic and dangerous. Bring children and pets indoors, and follow the procedures laid out above for handling a nuclear-reactor emergency. Because CBR agents are heavier than air, move to the highest possible floor.

If you're indoors and a lethal agent has contaminated the interior of your building, evacuate and travel upwind.

To get out of harm's way, you need to reduce your exposure as much as possible if a CBR agent has been released in your building. This could have occurred as a result of an explosion or the poisoning of the building's ventilation system. Evacuate the building, taking pains to avoid or minimize passage through the contaminated area. Close doors behind you as you leave, along with windows, if possible. Keep going until you find emergency assistance.

🚫 Assuming you may have been contaminated, don't touch anyone — or let others touch you — except for trained emergency personnel. Various CBR agents can be transmitted by personal contact.

If you're outdoors and an outdoor CBR incident takes place in your immediate vicinity, seek refuge indoors.

Get clear of the source of the contamination as quickly as possible. If the air is saturated with a CBR agent, fleeing on foot along streets or driving down roads will only increase your exposure. Instead, seek refuge in a building, making sure it isn't at the epicenter of the incident. Move away from doors and windows. Go to the highest floor and find, if possible, a windowless inner room. Most CBR agents are heavier than air and sink over time, making higher floors safer than lower ones.

If you can only find rooms with windows, make sure none of the windows were open during or after the attack. Before entering the room—or before allowing anyone else in—discard any contaminated clothing. Strip to your underwear if you have to. Place your clothes in a bag or container and leave it outside the safe-haven room. Shut down any air-ventilation system, if possible. Tightly seal any windows and doors with plastic tape. Plug keyholes with cloth. Use cloth or paper also to fill gaps around and under doors. If heavy-duty plastic bags or sheets (preferably, six mil or more) are available, use them to cover the doors and windows.

 If you spot a bottled-water cooler, bring it into the room. You may need it.

Put barriers between you and the attack site.

It can be helpful to try to put physical barriers between you and the site of a CBR incident. These might be natural barriers, such as hills, mountains, and dense forests, or man-made ones, such as tall or large buildings—or even vehicles if you're near the epicenter of the attack. The barriers will act as a shield, potentially blocking your exposure to lethal agents.

Wash as soon as possible if splattered with a CBR agent.

If no water is available and the agent was a liquid, use a powder like flour or talc to dry off the affected skin. Brush off the residue, being extremely careful not to inhale any of the dust. Be careful how you discard any towels or cloth you use.

As soon as possible, discard the clothes you've been wearing and shower.

Once clear of the contaminated area, remove all external apparel (i.e., clothes, shoes, gloves, and hats) and leave them in a sealed bag outside. Proceed to a shower, thoroughly washing your body with soap and water. Showering needs to be accomplished as soon as possible after a CBR attack. But don't simply flush water over your body. Give yourself a really good scrubbing, preferably with antibacterial soap. Scrub your skin aggressively and irrigate your eyes with water.

Follow the instructions of national emergency response teams.

A national emergency response plan, involving federal, state, and local agencies, will be activated if a large quantity of biological/chemical agents or radiation is released, and the public will be advised to take basic protective measures. As an example what to expect, consider the following verbatim advice from the CDC on the steps to take in a radiological emergency:

- Seek shelter in a stable building and listen to local radio or television stations for national or local emergency-alert information.
- Follow the protective-action recommendations from state or local health departments. Reduce your potential exposure and adverse health consequences by getting away from the radiation source, increasing your distance from the source, or keeping behind a physical barrier such as the wall of a building.
- If an event involves a nuclear power plant, a national emergency response that has been planned and rehearsed by local, state, and federal agencies for more than 20 years will be initiated. If you live near a nuclear power plant and have not received information that describes the emergency plan for that facility, contact the plant and ask for a copy of that information. You and your family should study the plan and be prepared to follow instructions from local and state public health officials.

"Local authorities will issue public health and safety statements advising precautions to take to avoid potential exposure to radiation," the CDC adds. "Until the amount of contamination is determined, the following precautionary measures are recommended to minimize risk:

- Remain inside and avoid opening doors and windows.
- Keep children indoors.
- Turn off fans, air conditioners, and forced-air heating units that bring in fresh air from the outside. Use them only to recirculate air already in the building.
- Go to the nearest building if you are outside. If you must go outside for critical or lifesaving activities, cover your nose and mouth and avoid

stirring up and breathing any dust. Remember that your going outside could increase your exposure and possibly spread contamination to others.

- Be aware that trained monitoring teams will be moving through the area wearing special protective clothing and equipment to determine the extent of possible contamination. These teams will wear protective gear as a precaution and not as an indication of the risks to those indoors.
- Avoid eating fruits and vegetables grown in the area until their safety is determined.

✛ CHAPTER 3 ✛

Cyberterrorism and Internet Security

I t's estimated that some 58,000 known viruses and other computer security threats lurk on the Internet—and the number is rising daily. Computer experts warn that some of these threats could erase your entire hard drive.

And don't think that cybervandalism can't affect you. If your computer is hooked up to the Internet or if you download material from floppy disks or CDs and you don't take the proper precautions, you could someday find your system plagued by viruses or your stored information deleted or mangled.

Then, too, there's the danger of cyberterrorism. Most virus writers and hackers these days seem to be adolescents looking to make mischief. They often are too young and inexperienced to fully appreciate the immensity of the damage their viruses do and the amount of time and experience people devote to PC protection when they could be doing other, more productive and profitable things. Imagine, therefore, what might happen if terrorists made a determined effort to engage in cybercrime and destruction. The results could be catastrophic.

PC AND PASSWORD PROTECTION

Protect your personal computers and handheld devices with antivirus software and firewalls.

As advanced economies like that of the United States become increasingly information-dependent, cyberterrorism will become an ever-increasing danger. In early January 2002, experts discovered an especially vicious computer worm called JS.Gigger.A@mm. Spread via e-mail, it tries to delete all your files and make it impossible for you to restart your computer. This is just one example of a growing Internet security problem. Malicious viruses, worms, and Trojan horses are being sent around the world via infected e-mail, contaminated Web sites, and pinging programs at such a fast rate that antivirus software providers now sometimes update their lists of known viruses daily (as opposed to weekly). Some infected e-mail doesn't even have to be opened to activate; the W32.Badtrans.B@mm worm is an example of this.

Hackers also have their computers programmed to scan all online Internet users, looking for a vulnerable computer they can enter and exploit using a backdoor technique. Palukka is one. This backdoor Trojan horse gives a hacker access to your computer. Terrorists could employ Trojan horses, embedded in your computer and thousands of others, to execute denial-of-service attacks on any Web site in the world. They could also delete or retrieve your files or disable your computer.

It's imperative, therefore, that you install reputable antivirus software on all your computer equipment and erect firewalls as well. Then, you must discipline yourself to keep these updated continuously.

✔ Some antivirus and firewall programs now feature "live" or automatic updates, meaning you don't have to remember to do anything. The software does it for you.

✔ Good sources for the latest news on new viruses, worms, Trojan horses, and hackers include Carnegie Mellon University's CERT Coordination Center (www.cert.org), the National Infrastructure Protection Center (www.nipc.gov), Symantec Corp. (www.symantec.com), and the U.S. Department of Justice's cybercrime Web site (www.cybercrime.gov).

Test your computer's vulnerability to online security threats.

If you think your computer is safe, think again. PCs are more open to outside threats than you likely imagine. Indeed, the people who want to steal your personal and financial information are nothing if not ingenious. They seem to come up with new ways of penetrating your system almost every week. It's therefore essential to check your computer for vulnerabilities to online intrusion. Symantec (www.symantec.com), for example, offers a free and easy online tester to check your computer's vulnerability to online infiltration. It will run a program to determine how easy it would be to seize control of your computer and access sensitive information, including financial information and passwords. The results may surprise you.

✔ **Check out Symantex's latest virus threat news at http://securityresponse.symantec.com/. The site offers such interesting tidbits as the dates each virus was discovered.**

Make it a habit to update your computer software regularly.

Software manufacturers like Microsoft (www.microsoft.com) have automated facilities on their Web sites through which you can download updates, patches, and fixes to correct security weaknesses in your computer's operating system and programs. It's amazing how often these security problems crop up. Unless you keep your software continually updated, sooner or later a malicious program will wreak havoc with your computer. Contact your computer's manufacturer for assistance.

Be protective of your online names and passwords, and don't pick ones that are easy to crack.

Too often people use the same name and password to access a variety of Internet sites. These sites might range from the registration-required site of the *New York Times* to an online bank or brokerage. The danger, of course, is that

Most Commonly Used Internet and PC Passwords in Britain, 2002

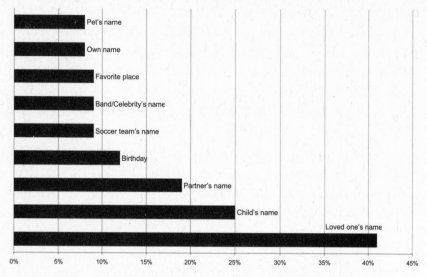

Source: Egg plc

if someone gets your name and password, they could possibly trade on your accounts or even steal your money. Make it a point to use different names and passwords for different sites. Don't use anything as simple as "bank1" or "IRA02" or nicknames like "Abby" or "Bill" that a hacker could deduce from your given name. Also, never use your birthday, your own name, or your Social Security number. An identity thief could use that information to establish fraudulent accounts or take out loans in your name. Interrupt any string of letters, as in a name, by interspersing a few numbers. Try using the first letters of a phrase or saying, such as TFTATFT or T42N24T as in "Tea for two and two for tea."

✔ Mine the names of nonimmediate family members (e.g., aunts, uncles, cousins, nieces, nephews, grandparents, etc.) or family friends for possible names and passwords with interspersed numbers that are easy for you to remember but hard for hackers to figure out.

Avoid using your pet's name, your own name, or the name of a spouse or child. Also, avoid birthdays.

Be especially mindful of what you store on your portable computer.

Laptops are great, but unfortunately they also get stolen. So be protective. Think twice about putting sensitive personal data and information on your laptop. And if you must carry around critically important information on your computer, use what's called a strong password, composed of a combination of letters—both upper- and lowercase—numbers, and symbols.

THWARTING INTERNET INVADERS

In August 2003, three new Category 4 worms hit the Internet within less than two weeks. The Blaster, Welchia, and Sobig.F worms infected millions of computers worldwide and inflicted as much as $2 billion in damages.

Think twice whenever you're online.

In conjunction with its effort to curb the tide of identity theft, the Federal Trade Commission makes the following suggestions on how to protect your computer and the personal information it stores:

- Don't download files from strangers or click on hyperlinks from people you don't know. Opening a file could expose your system to a computer virus or a program that could hijack your modem.
- Use a secure browser—software that encrypts or scrambles information you send over the Internet—to guard the safety of your online transactions. When you're submitting information, look for the "lock" icon on the status bar. It's a symbol that your information is secure during transmission.
- Avoid using an automatic log-in feature that saves your user name and password; and always log off when you're finished. If your laptop gets stolen, the thief will have a hard time accessing sensitive information.

For more on ID theft, see chapter 7, "Protecting against Fraud and Identity Theft."

Confirm bona fides when conducting financial transactions on the Web.

Appearances can be deceiving, especially on the Internet. So when conducting financial affairs on the Web, make sure that the companies you're dealing with are bona fide.

For more information, see chapter 7, "Protecting against Fraud and Identity Theft."

Erase sensitive information on your hard drive before disposing of a computer.

PCs are so inexpensive these days that it's often cheaper and easier to buy a new one than to fix an old one. But what about disposing of your old computer? Be sure to delete any personal information stored on your computer before you put it in the trash.

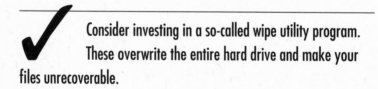

Consider investing in a so-called wipe utility program. These overwrite the entire hard drive and make your files unrecoverable.

Have two different Internet service providers.

A terrorist attack on critical infrastructure could cause telephones to go dead. E-mail could thus become a vital communications link, as happened in New York City on September 11 when the phone lines were jammed. However, if your home computer accesses the Internet via a telephone hookup, you'll be out of luck. Alternatively, there's no certainty that other Internet service providers (ISPs) won't be affected, too. To be on the safe side, subscribe to two different ISPs, if possible, and make sure that no more than one of them is linked via your home telephone.

INFORMATION SYSTEMS SECURITY

Information technology (IT) is the lifeblood of modern economies, but as everyone knows, computers and Internet communications are extremely vulnerable to viruses, worms, Trojan horses, denial-of-service attacks, tampering, and the like. This combination of economic importance and systems vulnerability makes IT a tempting target for terrorists.

Be sure employees follow proper security protocols.

IT security precautions only work if people use them. So be sure that employees follow the proper security protocols, such as password use and the protection of portable computers while traveling. Remind employees to check any new software for viruses before installation and not to store passwords on PF keys. Create an automatic log-off after a computer has gone unused for more than 10 minutes. Distribute, for example, the brochure "Computer Security Awareness," published online by the New York State Office of Technology at www.oft.state.ny.us/security/csa.html, and the Chicago Federal Reserve Bank's "Money $mart" guide to "Safekeeping Web Profiles and Passwords" at www.chicagofed.org.

Conduct a security check of your IT systems.

The Computer Security Division of the National Institute of Standards and Technology (NIST), a unit of the U.S. Commerce Department, has published the "Security Self-Assessment Guide for Information Technology Systems" to evaluate the security of a particular computer system or group of systems. Although intended for use by government agencies, the guide is a useful tutorial in IT security checks. "Through interpretation of the questionnaire results," it says, "users are able to assess the information technology (IT) security posture for any number of systems within their organization and, in particular, assess the status of the organization's security program plan." The guide is available at http://csrc.nist.gov/asset/.

✓ Carnegie Mellon University's CERT (Computer Emergency Response Team) Software Engineering Institute has published a guide to help organizations improve the security of their networked computer systems. See "CERT® Security Improvement Modules" at www.cert.org/security-improvement/.

Make a risk assessment of your IT systems.

IT risk assessment is one of the proactive security measures businesses need to take in this new age of terrorism. The NIST explains how to gauge the likelihood of an IT attack and the resulting impact on an organization in its "Risk Management Guide for Information Technology Systems." The guide outlines a nine-step procedure to conduct a proper IT risk assessment—specifically, system characterization, threat identification, vulnerability identification, control analysis, likelihood determination, impact analysis, risk determination, control recommendations, and results documentation. You can download the guide at http://csrc.nist.gov/publications/nistpubs/index.html. For further information, call 301-975-2934 or 301-975-6478.

Develop the means to detect IT intruders.

According to the Computer Security Institute's "2002 Computer Crime and Security Survey," 90 percent of respondents—mainly large corporations and government agencies—detected computer security breaches in the last 12 months, and 80 percent acknowledged a resulting financial loss. Nearly $456 million in losses were reported. Clearly, detecting intruders, such as hackers, is important to avoid unnecessary costs and to maintain the integrity of your business data and other proprietary information. Have qualified professionals install an intrusion detection system to stop trouble before it begins. For a further discussion of this topic, see the papers on "cyberstrategies" presented at a New York State Security Conference, April 10–11, 2002, which are located at www.oft.state.ny.us/security/2002secday.htm.

Encrypt sensitive data, especially customers' credit card numbers.

Practice IT defense in depth by employing firewalls and backup strategies.

Firewalls are essential these days, given the proliferation of infected e-mails and Web sites. But choosing, configuring, and maintaining firewalls for a business can be difficult. The NIST, in January 2002, published a set of important recommendations in its "Guidelines on Firewalls and Firewall Policy," available at http://csrc.nist.gov/publications/nistpubs/index.html.

The NIST also issued this key warning: "Firewalls are vulnerable themselves to misconfiguration and failures to apply needed patches or other security enhancements. Accordingly, firewall configuration and administration must be performed carefully and organizations should also stay current on new vulnerabilities and incidents. While a firewall is an organization's first line of defense, organizations should practice a defense-in-depth strategy, in which layers of firewalls and other security systems are used throughout the network. Most importantly, organizations should strive to maintain all systems in a secure manner and not depend solely on the firewalls to stop security threats. Organizations need backup plans in case the firewall fails."

Stay abreast of the latest IT vulnerabilities and incidents.

Keeping on top of IT vulnerabilities, new viruses, and IT attacks is a vital yet relatively easy chore. See the "Cyberterrorism and Computer/Internet/Information Security" section of the appendix.

Provide a means of communicating security alerts throughout your company.

Knowledge of a security threat can reduce its impact on a company, so have a mechanism in place to send security alerts instantaneously throughout your enterprise. Similarly, standardize the way employees report incidents to your IT security staff.

Explain the importance of IT security to new hires and office temporaries.

Ensure your IT security personnel are properly trained.

IT security is an ever-changing challenge, given the new threats that arise weekly, if not daily. It's imperative, therefore, to have well-trained IT professionals overseeing your systems—and for them to keep their training up-to-date. The New York State Office of Technology (www.oft.state.ny.us/security/security.htm) cites the following computer-security training programs: SANS Institute, 866-570-9927 or 540-372-7066, www.sans.org; MIS Training Institute, 508-879-7999, www.misti.com; Computer Security Institute, 415-947-6320, www.gocsi.com; TruSecure's International Computer Security Association, 888-627-2281 or 703-480-8200, www.icsa.net; Information Systems Security Association, Inc., 800-370-4772 or 414-768-8000, www.issa-intl.org; and International Information Systems Security Certification Consortium, Inc., www.isc2.org.

Be wary of outside IT contractors.

Whenever anyone from outside your company works on your IT systems, you're open to malicious or criminal activity. It's essential, therefore, that you hire only reputable contractors with impeccable credentials. In congressional testimony in 1999, Michael A. Vartis, director of the FBI's National Infrastructure Protection Center, listed a host of dangers. Consider these:

(1) *Systems maps:* By mapping your IT systems, an outside contractor could gain valuable information to sell to the highest bidder.

(2) *Root access:* Contractors are often given the same access privileges as the systems administrator, allowing them to steal or alter information or engage in a denial-of-service attack.

(3) *Trapdoors:* By installing trapdoors, contractors could gain access to your systems at a later date through openings they created.

(4) *Malicious code:* In writing code, someone could place a logic bomb or a time-delayed virus in a system that would later disrupt it.

(5) *Compromised security:* A malicious actor could implant a program to compromise passwords and other aspects of IT system security.

("Year 2000 Technology Problem," National Infrastructure Protection Center, Federal Bureau of Investigation, Statement for the Record before the Special Committee on the Year 2000 Technology Problem, U.S. Congress, July 29, 1999.)

✦ CHAPTER 4 ✦

Mail Handling Guidelines

While there have been scares and false alarms, the mail hasn't been targeted by bioterrorists since the post-9/11 attacks of 2001. The first inklings of a problem came on October 5 of that year, when a sixty-three-year-old photo editor for a Florida publisher died from inhalation anthrax. Anthrax spores were later found in the building in which he worked. By the end of November, two Washington postal workers, a woman in New York, and another in Connecticut also died of the rare disease. More than a dozen other cases of anthrax exposure were eventually confirmed, and anthrax was discovered in letters sent to journalists in New York and to members of Congress. No one has ever claimed responsibility for the lethal letters.

THE SAFEST WAY TO HANDLE MAIL

Amazingly, long after the anthrax letter attacks of 2001, government authorities still couldn't agree on exactly how people should handle their mail. We've reviewed a number of sources of information, including the U.S. Postal Service (www.usps.com), the Centers for Disease Control and Prevention (www.bt.cdc.gov), the Los Angeles County Department of Health Services (http://labt.org), and the British Government Co-ordination Centre (www.co-ordination.gov.uk). The recommendations below incorporate all of the suggestions made by these organizations and are augmented by our own understanding of the bioweapons threat. These suggested procedures offer the most thorough safeguards formulated to date. Let's go through the process step by step.

Know how poisoned mail can harm you.

Biological agents (as discussed in chapter 2 on the nature of the threat) have to get into or onto your body to do you harm. They usually do this in three ways: (1) inhalation: you breathe in an airborne biological agent, and it enters your lungs; (2) ingestion: you swallow the agent, and it gets into your digestive system; (3) contact: you touch the agent or it comes in contact with your skin. Another, less frequent way is through an open wound or cut that permits the agent to enter the bloodstream. So to reduce the risk of biologically tainted mail infecting you, you want to avoid inhaling or ingesting the agent or letting it linger on your skin or enter through broken skin. The following steps aim to accomplish that.

 Don't eat, drink, or smoke around mail.

Begin by creating a regimented routine for handling your mail and stick to it.

Gone are the days when you can toss mail around willy-nilly. Treat your mail like you treat fresh meat, fish, or poultry after bringing it home. Normally, you don't put it down just anywhere. No, you put it in the refrigerator until you're ready to cook it. Well, think of your mail in the same way. Always put it in the same place until you're ready to open it. Select an isolated spot in your home (or office) that's away from food. Use the same location all the time. When you do sit down to open your mail, do it conscientiously. Always use a letter opener. Don't just tear your mail open with your fingers, because that could disturb any malicious contents, contaminate your skin, and even cause a paper cut. Never ever blow into an envelope to open it. Don't shake letters or parcels. Don't pour out any powdery contents. Try to open your mail with a minimum amount of movement and disturbance to its contents. Keep your hands away from your nose, eyes, and mouth while handling mail.

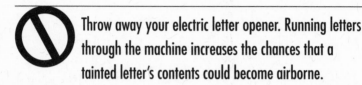

In a multiperson household, it might be wise to designate one person as the primary mail opener. That's not to say that every member of your household shouldn't know the proper way of opening mail. But by designating a person to handle the chore, your family's anxiety will be lowered and the chances of a mishap lessened.

🚫 Throw away your electric letter opener. Running letters through the machine increases the chances that a tainted letter's contents could become airborne.

Make it a standard practice to wash with antibacterial soap immediately after touching mail.

Thoroughly washing your hands and wrists for at least 15 seconds with antibacterial soap can rid you of exposure to such germs as anthrax. Make this a habit. Remember that terrorists use surprise as a means of spreading fear. When the next lethal-letter campaign begins, you won't be given advance warning.

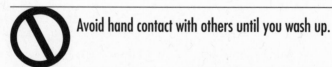

Avoid hand contact with others until you wash up.

Promptly discard all empty envelopes and parcels, and decontaminate the area.

Biological cross-contamination through the mails remains an ever-present danger—and one that foreign terrorists may decide to exploit in the future.

All envelopes and emptied parcels should thus be discarded immediately after opening. Don't let the contents come in contact with the outside of envelopes or packaging after opening. Place all mail contents in a special, uncontaminated spot. Discard envelopes after opening by placing them in disposable plastic bags other than ones used for normal trash. Before reading your mail, decontaminate the surface area by moistening a paper towel with an antibacterial disinfectant and wiping the tabletop and any trays used to hold the mail. Don't spray the disinfectant onto the surfaces, because that could cause deadly microbes to fly into the air. Discard the used towels in the plastic bag and seal it. When closing an envelope-filled bag, keep the opening away from your face, trying to avoid inhaling any emitted dust or getting any microscopic debris in your eyes.

 Disinfect your letter opener and anything else you may have touched in opening the letters, such as a lamp or telephone.

 If you have cuts on your hands or broken cuticles, wear latex or other protective gloves when handling mail or parcels, as biological agents can enter the bloodstream through broken skin.

Wear protective gear if it gives you peace of mind.

There are no hard-and-fast rules about wearing protective gloves, masks, and eyewear while opening mail. It would be safer if you did, but taking sensible precautions in determining whether a piece of mail is suspicious and following the instructions provided above should be sufficient, unless some new, major outbreak of a mail-borne illness occurs. Still, if it brings peace of mind, outfit yourself at a local medical supply store with disposal gloves, masks, and eye protection. Some officials have pooh-poohed the wearing of masks, etc., for what appears to be political rather than health reasons. Treat whatever government authorities say on this matter with circumspection. Erring on the side of caution has never killed anyone. Anthrax-tainted letters have.

DETECTING LETHAL LETTERS AND PARCELS

International terrorist organizations haven't shown much interest in sending bombs through the mail (as of yet, that is), but domestic terrorists in the United States have. Today, around 200 groups in the United States are known to be actively seeking forcible political change. The Los Angeles Police Department reports that its bomb squad responds to about 900 bomb-related calls a year, of which 35 percent actually involve explosive materials, including dangerous fireworks. Moreover, according to federal authorities, most U.S. bombers deliver the devices themselves.

Carefully inspect your mail before opening, and know the telltale signs of suspicious letters and parcels.

Begin by scrutinizing your mail for any letters or parcels sent by an unfamiliar party.

Take a careful look at items that seem suspect, paying close attention to postage, postmarks, labeling, and appearance, including size, shape, and weight. The signs below cover mail that could contain a biological agent, a toxic chemical, or a bomb.

POSTAGE   Look for excessive postage, no postage, or noncanceled postage.

POSTMARK   Cross-check all labels against the stamped postmarks to see if they match. Be suspect of unexpected mail from a foreign country.

LABELING   Be wary of handwritten or poorly typed labels from unknown sources, as well as any mail addressed using purposely distorted handwriting or cut-and-paste lettering. Be cautious of mail that has no return address or that clearly has a fictitious return address. See if your name is misspelled, your job title is incorrect, or your address is wrong. Be careful of mail addressed to you only by your job title and not your name, or mail addressed to someone no longer with your company. Finally, be suspect of any mail with odd restrictions or special handling instructions (e.g., "For Your Eyes Only," "RUSH," "Don't Delay," "Handle With Care," etc.) or a threatening message scrawled on it.

APPEARANCE Look for any crystals or powder on the surface, discoloration of the wrapping or envelope, oily spots, or stains. If an item has an unusual odor, stay away from it. Visually inspect suspect items for protruding wires, aluminum foil, soft spots, or ticking sounds. Get a general impression of a parcel by determining whether it has been sloppily wrapped, tied with an excessive amount of string, or heavily taped. Look for packages that are lopsided, bulging, uneven or lumpy, oddly rigid, or irregularly shaped.

FEEL Lift a suspect parcel to determine if it's excessively heavy for its size or lopsided with most of the weight on one side. This could indicate a bomb. Feel the item for any contents that may have settled to the bottom or corners. Powdery substances can often be felt through an envelope or package. Be careful, however, not to flex or bend a suspicious piece of mail.

✔ Letter bombs or other lethal items needn't arrive via the postal system or even a private courier service. Often they're simply left in a mailbox or on a doorstep by the bomber.

🚫 Never shake a suspect piece of mail. If it's a bomb, it could explode. If it contains a hazardous agent, shaking could spread the contents into the air. And never stick your nose into a suspicious package or envelope to sniff its contents.

Follow proper procedures if you encounter a suspicious letter or parcel.

If you're at home and discover suspicious mail, put the item down without shaking it or emptying its contents. Don't carry the package or envelope to show to others or allow others to touch it. Promptly leave the room, closing doors behind you. Turn off all radios and cell phones in the vicinity, because they could trigger a bomb, and shut off the air-ventilation system. Vacate the premises, and call the police from a landline phone. Avoid touching anyone until you wash with soap and water if you suspect a biological agent.

If you're at work and you come across a suspicious piece of mail, put the envelope or package in a plastic bag or other container to prevent leakage of its contents. If a bag or container isn't available, cover the suspect item with anything that's at hand (e.g., clothing, newspaper, a trash can, etc.). Leave the room, closing the door behind you, and tell workers in the area to vacate the building. Turn off radios and cell phones. Call building security or the police using a landline phone. Request that the building's air-conditioning unit be switched off, and close any open windows in adjacent rooms. Seal off the area around the suspect mail, and keep others from entering until help arrives.

If you suspect a biological device, wash with soap and water, and get medical attention as soon as possible. If help takes long to arrive, disinfect yourself by washing your hands and arms with a mixture of one part household bleach to 10 parts water, as this may reduce the possibility of absorbing an agent through your skin, but be careful not get the solution in your eyes.

After washing up, make a list of all the people who were in the room or area when you spotted the suspect letter or package. Provide the list to law enforcement, public-health officials, and your employer.

 Never bring suspicious mail to a police station, fire department, post office, or doctor's office. Leave it where it is, and let the experts handle it.

Don't be deceived by a package's size.

Letter bombs can range from the size of a cigarette pack to a large parcel. The contents of letter bombs are often stiff or springy. Never bend a suspect item of mail because that could set a bomb off. Timers may or may not be used to detonate the device. Most often, opening the parcel triggers the bomb. Again, listen for any ticking sounds and inspect for any exposed wires or aluminum foil. If you're the least bit concerned, call the police and leave the premises.

2

TRAVEL SAFETY

CHAPTER 5

Travel Guidelines and Concerns

Attitudes about travel, notably air travel, have changed considerably (and for the better) since the first weeks and months after 9/11. Apprehension and even fear gripped many fliers back then. Most everyone expected terrorists to at least attempt other jetliner takeovers, but this time many more passengers could be expected to join in the fight to thwart the hijackers. Nowadays, hijacking seems a less imminent threat. Improved airport and airline security, along with the larger effort to declaw terrorists globally, has certainly salved many passengers' frayed nerves.

Still, hijackings can never be ruled out completely. Keep in mind, too, the horror of the simultaneous bombings of commuter trains in Spain in 2004. What is more, travel to many parts of the world entails inherent risks simply because of high crime rates, as well as social unrest, political ferment, and economic distress.

Travel safety tips will never go out of style. Still, it's important to put the risk of foreign travel into perspective. Statistically, you're far more likely to become a victim of crime overseas than a victim of terrorism. In that sense, the dangers of traveling at home or abroad aren't all that much greater than they used to be—at least when it comes to the chances that any one individual traveler will fall victim to violence. Still, the threat of terrorism cannot be ignored. Americans will be targeted. So what's required is a redoubling of our efforts to stay safe.

Be willing to change and receptive to new ideas.

Appreciate first that you can't simply will yourself to be safe. Travel security requires an investment of time and effort. Reading this chapter is a start—but only a start. Next, be willing to change. Your old travel habits are not appropriate in this new age of terrorism. Finally, be receptive to new ideas and new ways of doing things. Approach travel security with an open mind, and jettison your old notions about safety. Don't reject suggestions out of hand or let preconceptions get in the way, because most assumptions about travel safety formed over years of experience are no longer valid.

GETTING INFORMED/STAYING CONNECTED

Information is knowledge, and knowledge is power. The better informed you are about the nature of the terrorist threat as it pertains to your travel plans, the safer you will be. Don't look at the task of acquiring travel-related information as a burden; think of it as an investment.

Familiarize yourself with your travel destination by reading the local newspapers and listening to local radio news online.

Don't create an artificial separation between terrorism and crime. Either one can be lethal. So make it a point to know what's happening at your travel destination. But don't rely on your hometown newspaper to give that information. Instead, go online and read the local papers and listen to local radio news in your travel city—especially one you're visiting for the first time.

We recommend three directories of local newspapers links: Kidon Media Links (www.kidon.com/media-link/index.shtml), even though based in the Netherlands, has links for virtually every newspaper in the United States (as well as the world). You can also find newspaper links at NewsLink (http:// newslink.org) and NewsDirectory.com (www.newsdirectory.com). For the most complete list of radio stations with online feeds, go to Radio-Locator (www.radio-locator.com).

Ask friends and colleagues for insights about your travel destination.

Reading a newspaper or listening to radio news can only tell you so much about potential risks of crime or terrorism at your travel city. Get as much information as you can from people who live there or have been there recently, such as coworkers, employees of your company based at your destination, clients you plan to meet, friends, or relatives.

✓ Strike up a conversation with passengers next to you on your outbound flight. If they live or work at your travel destination, they, too, might offer some invaluable insights, especially about the parts of town to avoid.

Avoid long waits at U.S. border crossings.

Given the potential for injury at U.S. border crossings were, say, a car bomb to go off or a gunfight to break out between terrorists and border guards, the less time spent waiting in line, the better off you'll be. For border-crossing wait times, see the U.S. Customs Service's Web site, www.customs .gov. Click "Travel Information" or go directly to www.customs.gov/travel/ travel.htm.

Avoiding trouble spots around the world is the surest means of surviving terrorism. So get in the habit of checking the latest official travel warnings.

You can hear the latest travel warnings and other recorded information about foreign destinations by calling the U.S. Department of State in Washington, D.C., at 202-647-5225. You can also get that information via automated telefax by dialing 202-647-3000 from your fax machine. Or you can go to the State Department's Web site at www.state.gov. In addition to warnings about travel to individual countries, the State Department also issues regional and worldwide cautions. On July 1, 2002, for instance, it advised:

The U.S. Government continues to receive credible indications that extremist individuals are planning additional terrorist actions against U.S. interests. Such actions may be imminent and include suicide operations. We have no further information on specific targets, timing, or method of attack. We remind American citizens to remain vigilant with regard to their personal security and to exercise caution. Terrorist groups do not distinguish between official and civilian targets. Recent attacks on worshippers at a church and synagogue underline the growing possibility that as security is increased at official U.S. facilities, terrorists and their sympathizers will seek softer targets. These may include facilities where Americans are generally known to congregate or visit, such as clubs, restaurants, and places of worship, schools or outdoor recreation events. Americans should increase their security awareness when they are at such locations, avoid them, or switch to other locations where Americans in large numbers generally do not congregate. American citizens may be targeted for kidnapping.

✔ The State Department's 202-647-5225 number can also be used by friends and families at home to get help in emergencies involving U.S. citizens overseas.

Get on the U.S. State Department's e-mail list for the latest terrorism advisories.

The U.S. State Department now offers a free e-mail distribution of terrorist-related news and information. It's a real boon for anyone planning a trip abroad or travelers already in a foreign country. The bulletins cover a wealth of subjects from new travel warnings and updated passport information to speeches by government officials and daily State Department press briefings. You can pick from the various categories to get just the e-mails you'd like. The list includes a special, nearly all-encompassing category called "America Responds: Building a Global Coalition Against Terrorism," which provides all of the State Department's announcements on the post-9/11 global antiterrorism effort. If you subscribe to that particular list, be aware that you can expect a substantial number of e-mails. Subscribing is as simple as can be. Go to www.state.gov/www/listservs_cms.html, make your selections, and type in your e-mail address. That's it. You're done. And if the e-mail traffic becomes too much or you no longer need the information, you can cancel your subscription to any or all of the lists by completing an online sign-off form.

Consult the international travel warnings issued by the British and Canadian governments.

Besides the U.S. State Department, the British Foreign and Commonwealth Office and the Canadian Department of Foreign Affairs and International Trade also issue timely, country-by-country travel warnings and other announcements. What makes these reports so interesting (and so valuable) is that they don't always agree with the U.S. State Department's assessment. Sometimes Britain or Canada (or both) will conclude that conditions in a foreign land are worse than the State Department is letting on, or they will provide more detailed information about the form and specific location of the threat. At other times, the British and Canadians may be more sanguine than the U.S. State Department. The British Web site is at www.fco.gov.uk, and the Canadian site is at www.voyage.gc.ca/destinations/menu_e.htm.

Don't be lulled into a false sense of security simply because you're on vacation.

Vacations are, of course, times to relax. That's why we take them after all. But the desire for rest and relaxation mustn't result in letting down your guard. Stay as alert on vacation as you are when traveling on business.

✓ Monitor the news for your foreign travel destination. Check the appendix for a complete list of foreign-newspaper and travel advisory contacts and Web sites.

Monitor the radio broadcasts of the Voice of America and the BBC World Service.

Be sure to keep your computer's antivirus software updated, and we strongly recommend that you invest in a firewall if you don't already have one.

Take advantage of the Internet's free translation services to read the foreign-language press.

While the Internet has plenty of news sources in English, you may find it necessary to read news that appears only in a foreign language. Finding foreign-language newspapers, wire services, and magazines on the Internet isn't a problem. This trouble comes when you don't know the language. Well, now there's a solution.

Foreign languages are no longer the barriers to information they used to be, thanks to automated, online translation services. You can as easily read a newspaper written in, say, French or German as in English. What's more, some of these online translation services are free! Our favorite free online translator is Alta Vista's Babel Fish (http://world.altavista.com). It will translate into English bits of text or even entire Web pages from Chinese, French, German, Italian, Japanese, Korean, Spanish, Portuguese, and Russian. While the quality of the translations wouldn't get you an A in school, the results are reasonably accurate and certainly sufficient to give you the gist of a story. Directions for the translator are simple and straightforward. And again, once you have the resulting URL, you can save it as a "favorite" and go back to it at any time.

Know where to find links to the foreign-language press.

There are several places to find the Web addresses for foreign-language newspapers. The best place to go, in our opinion, is Kidon Media Links (www.kidon.com/media-link/index.shtml) of the Netherlands, which has links for virtually every newspaper and wire service in the world. We do not mean hundreds of links . . . or even thousands. We mean tens of thousands of media links, including magazines and TV and radio stations. Kidon also tells you the language of a publication and how often it's published. Another worthwhile site is NewsDirectory.com (www.newsdirectory.com). You may also want to check out Radio-Locator (www.radio-locator.com), which has links to more than 10,000 radio stations around the world. Unfortunately, there are no free online translators for the spoken word.

Consider a country's economic condition when assessing the risk of going there.

In assessing the potential risk of violent street demonstrations or individual acts of violent crime, you must take into account a country's economic circumstances. A major currency devaluation, a huge rise in inflation and unemployment, a dramatic downturn in an economy—any of these could spark rioting, particularly in an already poor country. Street riots, for example, followed the recent money crisis in Argentina and the devaluation of the Turkish lira. Worse, Mexico's peso crisis of 1994–95 so damaged the economy that many out-of-work Mexicans turned to crime. Conditions in Mexico City indeed became so bad that car drivers refused to stop for red lights or stop signs, flew over speed bumps, and never rolled down their windows for fear of carjacking, abduction, and murder. Thus, it probably wouldn't be wrong to say that the worse off a country is economically, the greater the likelihood of violent crime. As a foreigner traveling in an impoverished country, destitute locals might perceive you as a rich foreigner, someone who is likely to be carrying a lot of money and valuables. You might also be viewed as an important foreign business executive who would command a king's ransom if kidnapped.

Because new terrorist threats or incidents could affect Americans overseas, and conditions in a foreign country can change in an instant, you must keep on top of the latest news while traveling outside the United States. Here are some suggestions:

Bring a laptop and stay only in hotels that provide for Internet hookups in your room.

Keeping abreast of breaking news about terrorism or crime that could directly affect you becomes most important when you're actually overseas. So bring a laptop computer or other Internet-capable device with you on your trip, and stay only in hotels where you can make a direct connection to the Internet from your room. Also, make sure you can retrieve your e-mails while traveling. And store your laptop not in its original carrying case but in a common, nondescript bag or satchel that won't attract a thief. Avoid public spaces in hotels that permit guests to use computers to access the Internet and get e-mail. Terrorists know that foreigners congregate in these areas, and thus these spaces are potential targets of terrorist attack. Criminals, too, could enter the hotel's computers later to try to retrieve sensitive information or data concerning your financial accounts. Therefore, do all you computer-related work in the privacy of your room or at the local branch office of your company. And never leave your computer in your room when you go out. Not only is your computer valuable to a thief, but the data stored within could be used to perpetrate a financial fraud or identity theft. Either take your computer everywhere you go, or put it in the hotel safe—but not your room safe. Room safes aren't very safe at all.

✔ Be advised that telephone jacks in many parts of the world are incompatible with U.S. phone cords, so you'll need to buy a plug adapter. Also, U.S. Customs may claim that you bought your laptop overseas; if you want to avoid paying duty, carry a copy of your receipt.

🚫 If you bring a laptop computer to Russia, you could be forced to leave it there. Russia's State Customs Committee has stated that there are no restrictions on bringing laptop computers into the Russian Federation for personal use. The software, however, can be inspected upon departure. Worse, some computer equipment and software brought into Russia by foreign visitors have been confiscated because of the data they contain or the software's encryption technology. (Encryption, of course, is standard in many programs.) In addition, the importation and use of a Global Positioning System (GPS) and other radio electronic devices are subject to special rules and regulations in Russia. For more information on rules concerning computers and GPS devices, see the State Department's advisory at http://travel.state.gov/gps.html. You can also contact the State Customs Committee of the Russian Federation, Russia 107842 Moscow, 1A Komsomolskaya Place. Its main telephone is 7-095-975-4070. Or you can call 7-095-975-4095 to get clearance to use personal items.

Rent or buy a cellular telephone with international roaming that will function at your foreign destination.

Cellular telephone networks around the world are incompatible with U.S. wireless technology and radio frequencies, meaning most cell phones used in the United States probably won't work overseas. Most foreign countries use a digital platform called Global System for Mobile (GSM) communications. More than 500 million cell-phone owners in 162 countries, representing 70 percent of the world's wireless subscribers, use the GSM standard.

You can buy GSM phones in the United States from such makers as Ericsson and Nokia, including international service, but be careful because some phones won't work in Japan, Korea, and Latin America. Remember, too, that you need to be signed up with an international roaming service for a GSM phone to work. It's best, therefore, to contact your wireless service provider first and ask questions. You may already have international-roaming access. VoiceStream (www.voicestream.com) sells and rents world phones with international roaming. Cingular Wireless (www.cingular.com) also provides this service. To rent a world phone with international roaming, consult the list of cell-phone rental companies listed in the appendix of this book. But caveat emptor. Let the buyer beware.

The good news is that most of the world's modern economies have cellular phone systems that are more advanced than those in the United States. In parts of Asia, for example, you can use a cell to get a soft drink from a machine dispenser or pay for your dry cleaning. This may come as a surprise to many of us who think of the United States as the world's technology leader. While that's true in many instances, it's not the case with cell phones. And there's a reason for it. State-run telephone systems have long dominated much of the world's telecommunications, especially in Europe. These state enterprises had been so horribly run and so unaccountable that businesses typically had to wait months or even years to get new telephone lines installed; private residences needing a new phone line could expect only a slightly shorter wait.

It's no wonder, then, that when cellular-phone services emerged, consumers and businesses outside the United States beat a path to their door. With so much pent-up demand for telephone service around the world, the cell phone industry began growing by leaps and bounds, offering the latest in gadgetry, technology, and service. Moreover, in developing economies, such as China, where the landmass is huge, the installation of cellular technology made a lot more sense than stringing thousands of miles of telephone lines.

 Be sure to pack spare cell phone batteries.

🚫 In Russia, you have to obtain permission to bring in a cellular telephone. An agreement for service from a local cellular provider in Russia is required, according to the U.S. State Department. That agreement and a letter of guarantee to pay for the cellular service must be sent to Glavgossvyaznadzor (the State Inspectorate for Communications), along with a request for permission to import the telephone. Based on these documents, a certificate is issued. This procedure is reported to take two weeks. Without a certificate, no cellular telephone can be brought into the country, whether or not it is meant for use in Russia. Glavgossvyaznadzor can be reached at Russia 117909 Moscow, Second Spasnailovkovsky 6; telephone 7-095-238-6331 or fax 7-095-238-5102. For more information, see the State Department's advisory at http://travel.state.gov/gps.html. You can also send an inquiry by mail or via facsimile to Consular Section, American Embassy, 19/23 Novinskiy Bulvar, 123242 Moscow, Russia; fax 011-7-095-728-5358. Inquiries from the United States can be sent to Consular Section, AM/EM - PSC - 77, APO AE 09721.

Beware of wiretapped phones and intelligence debriefings.

Foreign intelligence agencies are known to tap the phones of important foreign visitors, especially business executives. They may also debrief the people you meet with to discuss business—and even local hires at your branch office. So be careful with trade secrets and other proprietary information.

AVOIDING DANGER

With terrorists and criminals on the watch for foreigners, notably Americans, you need to take precautions in almost everything you do when traveling abroad. Here are suggestions on how to avoid some common mistakes:

Don't exchange currency at the airport.

Criminals often target foreign visitors exchanging large sums of currency at airport banks and foreign-exchange kiosks. If you need foreign currency for a taxi ride from the airport to your hotel, call the banks in your hometown before you leave to find one that will sell you the foreign currency you need. Otherwise, only exchange money at the hotel or a reputable bank.

 You could avoid the cost of a taxi ride by taking the hotel limousine from the airport.

Don't deal in the black market when exchanging currency.

Currency controls are still in force in some countries. This usually means that the official exchange rate, established by the national government, is a lot lower than the rate offered on the black market. U.S. dollars are highly prized around the world, especially in countries that have experienced ruinous currency devaluations. In those countries, the black-market trade in U.S. dollars is brisk. Indeed, don't be surprised if you're approached on the street by someone willing to exchange the local currency for dollars at a much better rate than the official one. Resist the temptation. First, you might be lured into a trap and robbed of all your valuables. Second, you might be given counterfeit money. Third, you might end up with a handful of old notes that have been taken out of circulation and are essentially worthless. Finally, you could land in jail. Buying and selling currency on the black market is illegal in many countries. For information on currency laws, see the U.S. State Department's individual country warnings and announcements at http://travel.state.gov/travel_warnings.html.

Hide maps and guidebooks when driving overseas.

When driving overseas, don't display signs of your foreignness by having English-language maps and travel guides in plain view in your car, especially when getting fuel or picking up food. Bury those items under a piece of clothing. The same goes for English-language newspapers, such as the *International Herald Tribune* or the *Wall Street Journal*.

Try to avoid large, self-park garages.

Parking garages are haunts for attackers, especially overseas. Therefore, use valet parking whenever you can, such as at restaurants and hotels, and try to avoid multilevel self-parking garages. Whenever possible, park your car in the open and in a well-lighted area visible to passersby.

Arrange your itinerary so as to drive in the daytime.

Don't start a long drive so late in the day that it will keep you on the road into the night. Try to drive only in the daytime when on a lengthy excursion. If you're going to encounter trouble while driving overseas, it's most likely to occur at night.

 Don't rely on the auto coverage offered by your credit card company; it usually doesn't cover such items as overseas medical expenses or personal liability.

In general, stay at U.S. hotel chains when abroad.

"Isn't it asking for trouble to stay at an American hotel?" you may be asking yourself. "Aren't American hotels the most likely targets of terrorism?" Well, yes, in some countries that may be true. In places where the risk that Americans will be singled out as terrorist targets is high, staying at a U.S. hotel chain may not be the most sensible thing to do. However, if the risk is that high, it's probably not wise to travel to the country at all until the threat subsides.

In reality, the greatest risks you'll face when traveling abroad will concern violent crime and food and health dangers. Security at U.S. chains tends to be much better than average. This lessens the chance that you could fall victim to burglary, assault, abduction, or murder while at your hotel. U.S.-owned hotels also conduct more sophisticated employee background checks and have better door locks and fire-safety regimes. The personnel background checks are especially important. Unless you can trust the hotel staff, you could be vulnerable to theft or worse. Indeed, rogue hotel employees have been known to provide information about guests to kidnappers and extortionists.

Less dramatic but no less risky is the danger of food poisoning. Sanitary procedures in the kitchens of U.S. hotel chains are a cut above average. In addition, U.S.-owned hotels tend to have qualified local doctors on call in case of health emergencies and also know which hospitals will provide you with the best care. Further, U.S. hotel chains typically have more modern communications facilities, making it easier to make calls, receive faxes, get on the Internet, and retrieve your e-mail. For Americans traveling overseas, U.S.-owned hotels also afford a natural affinity, meaning you'll probably be treated better and your safety and health will be a higher priority than at a foreign-owned hotel. Also, U.S. hotel chains are less likely to have secret surveillance cameras and listening devices in your room, planted by a state intelligence agency. But be aware that most hotels, pool areas, and bars tend be hangouts for pickpockets and other unsavory characters.

Finally, for Americans overseas, U.S.-based chains are more likely to agree to your special requests. You may, for instance, want the hotel to keep copies of vital information in case of an emergency or to contact the U.S. embassy if something goes tragically wrong. Managers of these hotels, while they may not be Americans themselves, must eventually answer for their conduct to their superiors in the United States. And the last thing a U.S. hotel chain wants is bad publicity. So odds are you'll receive the best treatment at hotels overseas that are part of U.S.-based chains.

Check with the regional security officer at the local U.S. embassy for a list of hotels used by U.S. officials visiting the area.

 If you're arriving after 6 p.m., be sure your hotel reservation is guaranteed; you don't want to be wandering around a strange city late at night looking for a room.

Think of the front desk at your hotel as "command central."

Take advantage of the security that your hotel itself offers. For example, make copies of your passport, visa, driver's license, health insurance cards, credit and bank cards, and prescriptions for drugs, eyeglasses, or contact lenses and place them in a securely sealed envelope. Then ask the front desk to put the package in the hotel safe. Having spare copies of your important documents and credit cards will make it a lot easier to get replacements if the originals are lost or stolen.

To help keep these private papers out of the wrong hands, affix a piece of strong tape or melt a bit of wax over the envelope's seal before handing the package over to the front desk for safekeeping. That way, you'll be able to tell afterward whether anyone has pried open your personal papers. Tell the hotel what you'd like them to do in case of an emergency and provide the names and telephone numbers, as well as e-mail addresses, of people they should call. If you have a serious medical condition or a life-threatening allergy—to penicillin or shellfish, for example—make the hotel aware of this, so they may inform a hospital or emergency medical technicians should the need arise.

In other words, think of your hotel's front desk as military command central, with you as a five-star general. Determine how the hotel can serve you, don't be shy about asking the front desk to do things for you, and give instructions on how you want them to handle certain situations. Also, leave a copy of your daily itinerary with the front desk, including scheduled meeting times, names, addresses, and phone numbers. Should you go missing, this information could prove invaluable to the police and even save your life.

> 🚫 We strongly recommend that you not hold these kinds of discussions in the open where an eavesdropper could hear you. Ask to be taken to a private office to discuss matters with the hotel manager. Also, try to develop a personal rapport with at least one senior staff member, whom you could rely on in a pinch.

Inform the local U.S. embassy of your presence and store vital documents there.

No matter how brief your stay in a foreign country—most especially one in which the risk of terrorism or other violence is high or if you are an executive with an internationally known company that could be a terrorist target— be sure to call the local U.S. embassy or consulate to inform them that you're in the country. See the appendix in this book for the phone numbers of U.S. embassies and consulates around the world. You can also get this information online from the U.S. State Department at http://usembassy.state.gov. The names of key U.S. diplomatic personnel around the world can be found at www.foia.state.gov/mms/KOH/keyofficers.asp.

Next, if you plan to stay more than a day or two, make copies of your passport, visa, driver's license, and health-insurance, bank, and credit cards and leave them with the embassy. And don't forget to retrieve these materials prior to departure. Inform the embassy, too, of any health conditions or serious allergies, and provide a list of people to contact in an emergency.

> Ask about the embassy's hours and where to go in an emergency.

Resist taking your spouse or children along at times of crisis or to high-risk destinations.

These aren't normal times, and many of the pleasures of international travel in the past no longer pertain. Traveling with your spouse or children is one of them. Be loving but be firm. Tell your family that it's neither the time nor the place for them to accompany you overseas. Make it up to them by taking them on a trip to a safer locale after you get back.

STREETWISE ADVICE

When in a foreign country, you're most vulnerable to attack when you're on the street. One reason is that terrorists and criminals usually canvass the streets, particularly near hotels and offices of foreign businesses, looking for targets of opportunity to kidnap or rob. Another reason is that foreigners tend to stick out on the street, especially if they make ostentatious displays of wealth.

Never take the first taxi in line.

If a group is out to kidnap you, one of the easiest ways is to gain your unwitting cooperation. Taxis are an especially useful method of deceiving an intended victim. A taxi might wait outside your hotel for your departure, or a compatriot might signal to the cabdriver that you're about to leave. To better ensure your safety, take the second or third taxi in line, but never the first one. Also, beware of unmarked cabs. Ask at the hotel desk for the names and phone numbers of reputable taxi, car, or limousine services.

 Never let anyone direct you to a specific taxi.

Vary your daily routines and routes.

Don't get into the habit of taking the same route at the same time every day from your hotel to, say, your branch office. Kidnappers and terrorists look for patterns. Vary your route and daily routine by asking the driver to take you by a specified site (e.g., a famous park, monument, or museum) before going on to your final destination. Also, leave and return to your hotel at different times of day. Furthermore, don't always eat at the same restaurant or go to the same tavern. Alternate any stores you might otherwise visit daily during your stay. If you're driving a rental car, park in different spaces, with the front of the car always pointing out, and keep the gas tank at least half-full to avoid being stranded or left searching for gasoline late at night.

 If you rent a car abroad, purchase the liability insurance; otherwise you could face a major and costly headache in the event of an accident.

Carry enough local coins to make a phone call.

International travelers commonly think of foreign coins as a nuisance. They can't generally be redeemed after you leave a country, so you're usually stuck with whatever you have in your pockets. However, don't eschew all foreign coins. You may need to make a local phone call in a hurry and your cell phone, for whatever reason, may not be working. If you're out of coins, you're out of luck. So make it a point to carry at least enough pocket change to call your hotel.

✔ Many international airports have charity boxes in which you can donate your leftover foreign coins. If you frequently travel to one foreign country, put your extra change in a marked envelope and store it next to your passport to use on your next trip.

Move away from public disturbances.

A loud disturbance on a public street is the kind of trouble you want to avoid. Don't be nosy. It could be a political protest about to turn violent or a terrorist-inspired incident. It could even be a distraction, used by pickpockets and purse snatchers to lift the wallets and purses of onlookers. Don't get involved in any street disturbance, and move away as quickly as possible in the opposite direction.

Don't advertise your nationality or foreignness.

Opportunistic criminals and terrorists look for foreigners. A good way to protect yourself abroad, therefore, is to look inconspicuous. You do that by appearing to blend in with the local population as much as possible. Of course, in certain areas of the world, that may be impossible. If you stand seven feet one like Los Angeles Lakers center Shaquille O'Neal, you're going to stand out on the streets of Tokyo no matter what you do. Still, there are ways to avoid sticking out like a sore thumb in a foreign land. First, never wear anything that gives your nationality away. You may be proud to be an American, for instance, but displaying an American flag pin on your lapel when traveling in, say, the Middle East might not be the smartest thing to do. Second, try not to wear clothing that automatically identifies you as a foreigner. That's not to say that you should take up wearing a turban when in India. That would only make you look foolish (and criminals eat fools for lunch). Instead, consider the cut and color of your clothing, and pick suits and outfits that most closely match those worn by the indigenous population. Finally, eschew wearing T-shirts, sweatshirts, or caps that identify you as an American.

Backpackers should resist the temptation to sew USA emblems and American flags on their packs.

Keep your voice down on the streets of high-risk cities.

If you're in a high-risk foreign country or city, speak as little as possible on city streets. Your native tongue or just the sound of your voice could be heard a mile away. If you must carry on a conversation, speak softly so as not to let strangers hear. Remember that eavesdropping is a tool used by criminals and terrorists to select victims.

Learn a few common phrases to get around a foreign city.

Not everyone has a gift for languages. Still, if you can learn enough common foreign phrases to get you where you're going, you'll give the appearance of being familiar with the territory. You may not pass for a native, but at least you won't broadcast that you're a complete neophyte. There's plenty of software available to help you learn foreign words and phrases, without incurring the time and expense of in-person lessons. Of course, nothing is as good as being taught a foreign language by a native speaker or an experienced teacher. However, for travelers who make only the occasional trip to a particular country, using a CD to learn a few helpful phrases may be just enough to get by.

Be suspicious of anyone who approaches you wanting to "practice English" or talk about the United States or politics of any kind; it's impossible to say these days if such an unsolicited approach is innocent or a ploy by a terrorist or kidnapper.

Always look as if you know where you're going, even when you're lost.

Give the impression of self-assurance wherever you go. Begin the moment you arrive overseas. Know precisely where your hotel is located, giving the taxi driver the name and street. But say it in a way that makes it sound as if you've been there before, even if you haven't. Mention the nearest cross street, perhaps. Follow the same procedure when asking a taxi to take you to a business meeting or restaurant. If you get lost, try not to show it. Never panic. Rather, go into a shop or café and ask for assistance. Otherwise, hail a cab or find the nearest phone and call a car service. Always book your hotel accommodations in advance. Never wander the streets looking for a place to stay. Do the same thing when dining out. Make reservations and know how to get to the restaurant before you leave your hotel. Sadly, much of the spontaneity traditionally associated with foreign travel died on September 11.

Don't telegraph that you're a stranger in town.

To lower your vulnerability in a foreign country, avoid doing the things that most first-time visitors do. For example, don't open up street maps or guidebooks in public. Try to memorize your route. Barring that, make a copy of the section of the map or guidebook you'll need that day. Glancing at what appears to be just a piece of paper is much less conspicuous than unfolding a huge map. If worse comes to worst, take a seat at a café, order a refreshment, and ask someone to point the way on your map. If you're a tourist rather than a business traveler, more than likely you'll want to see the sights—e.g., museums, art galleries, historic sites, etc. There's no need, however, to advertise that you're a foreign tourist. For example, if you want to go a famous museum by taxi, don't tell the driver, "Take us to the National Museum." Instead, give the driver nearby street coordinates and walk the rest of the way to the museum's entrance. Finally, if traveling to a high-risk area of the world, don't take your camera along. Standing on the street snapping pictures instantly tells a criminal or terrorist that you're from out of town. Buy photographs instead.

 Avoid pointing to things in public. You can always tell the tourists outside the Empire State Building; they're the ones with their fingers in the air.

Avoid public displays of wealth.

Displays of affluence make a criminal's heart grow fonder. You're only asking for trouble if you flaunt your wealth, especially in impoverished countries. Don't wear expensive jewelry or watches. And never take any family heirlooms abroad. Your rule of thumb should be, if you can't bear the thought of parting with a keepsake, leave it at home. Also, in some poor countries, displays of wealth can trigger hostility and resentment. As a foreigner, you could be subjected to some especially harsh treatment by the locals.

Keep any consumption of alcohol to a minimum.

It's vital to keep your wits about you these days, particularly overseas, so strictly limit any consumption of alcohol when traveling. And never drink and drive.

Don't keep all your credit cards and ID in one place.

Pickpockets are highly expert in many cities in the world. They are often far better trained, more experienced, and much more sly than anything experienced on, say, the streets of the United States. Pickpockets abroad are usually small in size and young. But don't let age, or even gender, fool you. Pickpockets come in all ages, sizes, and genders. But what they all share in common is the desire to get at your money, your valuables, and most especially, your credit cards. You can, of course, call your credit card companies to report a theft. But what do you do after that, particularly if you're overseas? The loss of your credit cards all at once could turn into a real nightmare. The smartest thing to do, therefore, is make it next to impossible to have every one of your cards stolen at the same time. Simply keep your credit cards in different places, so if your wallet is lifted, for instance, you'll still have a card in your pocket or briefcase to fall back on.

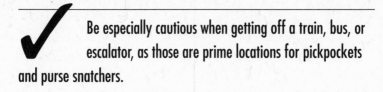

Be especially cautious when getting off a train, bus, or escalator, as those are prime locations for pickpockets and purse snatchers.

Be on guard against identity and passport theft.

Next to killing or injuring you, a terrorist would like nothing better than to steal your identity. Phony passports and identification cards are an essential part of a terrorist's tool kit. Buy special clothing or a body pack in which you can hide your identity papers.

Immediately report the theft or loss of your passport or credit cards to the proper authorities.

If your passport is lost or stolen, immediately inform the local U.S. embassy or consulate. You'll need to speak to the American Citizens Services unit of the Consular Section, where you'll be directed on how to obtain new passport photos. If you're about to leave the foreign country, provide the Consular Section with details of your departure schedule. Further information on lost or stolen passports is available at http://travel.state.gov/lost_passports_abroad.html. If your credit cards have been stolen, you'll need to contact your card providers and the U.S. Federal Trade Commission. The FTC serves as the federal clearinghouse for complaints by victims of identity theft. The FTC enters identity theft and fraud-related complaints into its Consumer Sentinel, a secure, online database available to hundreds of civil and criminal law enforcement agencies worldwide. The FTC can be reached at 877-FTC-HELP (877-382-4357) or online at www.ftc.gov. Its Web site contains an online form that you can use to report the theft of your identity. (For a more in-depth discussion of how to avoid identity theft and how to restore your credit should a theft occur, see chapter 7 on protecting your money and go to the FTC's Web site.)

EVERYDAY COMMUTING

Many Americans think they're really only vulnerable to terrorism when they're traveling by air on a long-distance trip for business or pleasure. Fact is, though, terrorists—notably in the Middle East but in Europe and Asia as well—often target commuters on buses and trains. What's more, bridges and tunnels used by commuters are also on the list of potential terrorist targets.

Keep an eye out for unattended items when using mass transit.

Because terrorists seek to kill and injure as many people as possible, mass transit offers a perfect target. That's why it's so important to notify authorities immediately of any unattended package, box, briefcase, luggage, or canister you may come across while commuting via train, subway, or bus. Unattended items could pose a serious threat and shouldn't lightly be dismissed. Similar caution should be taken in any public place where large numbers of people gather. Twelve Japanese commuters died and another 5,700 were injured when deadly sarin nerve gas was released into the air at a crowded Tokyo subway station in March 1995. A similar attack occurred almost simultaneously in the Yokohama subway system. Five members of the Aum Shinrikyo cult carried out the attacks, in which containers of gas were placed in subway stations and cars. Japanese courts have since sentenced several of the terrorists to death.

Learn how to spot suspicious behavior on streets, buses, and trains and at transportation stations.

Terrorists and criminals are often nervous and almost always look out of place in their surroundings. They don't behave like the others around and may be so preoccupied that they appear strange and unusual—as if something is pressing on their minds. They may even be jumpy. A group of men and even women who at first seem to know each other may suddenly act like strangers. Their actions may deviate from normal behavior, as if they were acting or imitating the everyday procedures followed by, say, janitors. They may be on the lookout for the police and may also pay far too much attention to something. Most often, the object of their attention is their intended target.

Note, too, that terrorists conduct reconnaissance missions prior to any attack, so you might see them photographing, videotaping, or mapping a location, testing out security doors or attempting to gain entry to areas that are offlimits to the general public. Spanish police in July 2002, for instance, arrested three Syrians suspected of being Al Qaeda members, and one of them had five videotapes containing images of New York's World Trade Center, Statue of Liberty, and Brooklyn Bridge, San Francisco's Golden Gate Bridge, Chicago's Sears Tower, and California's Disneyland and Universal Studios, as well as an unidentified New York airport.

Scout out the emergency exits at transportation stations you regularly use.

Time is precious in an emergency. You could do yourself a big favor in the event of a terrorist incident at a commuter station if you know in advance all of the available exits from the train, bus, or subway stations that you regularly use. Take some time and look around for the exits. Think about how you would get out of a specific location in a hurry. And don't consider just the most obvious routes of egress, because they could become jammed in an emergency. Scout out some of the less obvious exits, as well as places where you could take cover if, for example, gunfire were to break out.

> ✔ If you routinely use a major commuter hub that's a tempting target for terrorists (e.g., Boston's South Station or New York's Pennsylvania Station), consider changing your daily routine and taking an alternative route to and from work. You might, for instance, be able to catch your normal train but at a different station.

Ø Falling objects and flying glass cause many deaths and injuries in terrorist explosions. While waiting for your bus or train, therefore, make it a point not to stand under or near anything that could harm you in a blast.

Avoid rush hour by changing your commuting times, especially if you use major bridges or tunnels.

The crush of commuters at rush hour presents a tempting target for terrorists. That's because the types of devices they most commonly use (e.g., bombs) have their greatest effect in crowded places. Terrorists, therefore, look for venues where people are forced to be concentrated in large numbers or congregate voluntarily in mass gatherings. Crowded rush-hour train and bus stations are two such examples, but there are others. Terrorists, for instance, have set their sights on major bridges (as noted above) and tunnels. It's impossible to say in advance how well tunnels would hold up if a massive truck bomb exploded. It would depend on the age and construction of the tunnel and the size and location of the bomb. Most bridges would hold up rather well after the detonation of a car or truck bomb, with one exception. Bridges with two decks pose a special risk. The containment created between the upper and lower levels would magnify the force of a blast on the lower deck and could, under the right conditions, result in wholesale devastation and perhaps cataclysmic collapse.

Save for security checkpoints that might detect a bomb-laden vehicle, a commuter's only protection is to avoid the times at which terrorists would be most likely to attack. And those times are, of course, when bridges and tunnels are most heavily used—namely, rush hours. If possible, therefore, adjust your commuting schedule (or even your work hours) so as to avoid rush-hour traffic. The same rule holds for commuters using mass-transit trains, subways, and buses that cross bridges or go through tunnels.

✓ Talk with your company about telecommuting. You may be able to do work at home and reduce the number of days you have to commute to the office.

Never park in underground garages.

Terrorists already know that one way to cause mass destruction and maybe even bring down a building is to plant a car or truck bomb in an underground parking garage. So find another place to park that's outdoors. And when you do park, leave the front of your car pointing out so you can exit as quickly as possible in case a suspicious person is lurking about, and always check the interior of your car before getting in to see if an attacker is hiding in, say, the backseat. Also, park as close to an exit as possible. Consider asking for an escort to your car, especially when traveling at night.

SELECTING A HOTEL

Hotel selection requires more than merely considering cost, convenience, and comfort. Safety and security must also play a part in selection.

Make safety, not price, your prime consideration when selecting a hotel.

"That's easy for you to say. You don't know my company." If that was your immediate reaction to the above recommendation, you're working for the wrong company. Any firm willing to endanger you so it can save money may not deserve your loyalty.

 Check with the front desk to find out whether any of your hotel's entrances are locked at night; you don't want to be wandering about in the dark searching for an open door.

Don't just book the hotel nearest to your meetings or branch office.

Make the quality of the neighborhood in which your hotel is located a top consideration. It's tempting to pick the hotel nearest to your planned meetings or your company's branch office. And, indeed, these locations might be great in the daytime. But what about at night? Do they roll up the sidewalks? Do the streets become deserted? In most cases, it's best to find the safest area nearest to your points of call or your local office, then select your hotel accordingly. (We make an exception to this rule when it comes to foreign travel; our specific recommendations appear earlier in this chapter.) Find out, too, about the neighborhoods you plan to visit or pass through. Crime usually concentrates in different parts of a city like pools of water after a storm. Don't think that just because your hotel is in a safe area that the districts a few blocks away are equally tranquil.

✔ **Always carry your room key with you; don't leave it with the front desk, because thieves and kidnappers look in hotel mailboxes to see which guests aren't in their rooms.**

Avoid hotels that have underground garages.

Terrorists love underground garages. On February 26, 1993, Arab terrorists planted a massive truck bomb in the parking garage beneath the World Trade Center in New York City. The blast killed six persons and injured more than a thousand. Even though hotels have tightened garage security, there's little to stop a determined suicide bomber from driving through barricades and setting off an underground explosion that would likely cause the building above it to collapse. It's simply not worth the risk of staying on top of an underground garage. Find another hotel.

Make sure your hotel isn't a firetrap.

There's a fast and easy way to find out whether a hotel or motel in the United States—or even on Guam, the Northern Mariana Islands, Palau,

Puerto Rico, or the Virgin Islands—meets federal standards for fire and life safety. The U.S. Fire Administration (which is part of the Federal Emergency Management Agency) has an online database at www.usfa.fema.gov/hotel/search.cfm that you can search. If a hotel or motel makes the grade, you'll see a yellow "go" sign.

Don't hesitate to switch hotels if your needs aren't met or you sense something's wrong.

Hoteliers want your business, and therefore most are willing to accommodate any reasonable request. However, if you find yourself in a hotel that's unwilling to meet your needs, don't hesitate to switch to another one. Indeed, if for any reason you feel uncomfortable at the hotel in which you've been booked, move on. Your instincts are probably right. Whatever it is that you're feeling, odds are something is wrong with your accommodations. You may not be able to put your finger on it or put it into words, but when your sixth sense tells you that you're not safe, you probably aren't. Don't waste time trying to figure it out. It's not worth it. Call another hotel, pack up, and leave.

EXTRA SECURITY PRECAUTIONS

Personal safety in this age requires self-assertiveness. Travelers must become more demanding about their accommodations, for example, than they have been in the past. Not only will this help to better ensure your safety, but it also will raise awareness among hotel management that they need to pay more attention to counterterrorism measures.

Be particular about your hotel room and try to pick the floor you stay on.

Several considerations have to be weighed before accepting a hotel room. Assess the potential risks you face in that city. Is it a city afflicted with considerable crime? Is it known for terrorist incidents? When traveling overseas, weigh these additional concerns: Is there much public dissension with the government? Is there a history of arson or bombings of public buildings or attacks on foreigners? Then think about the room location that would best negate the more likely risks.

If kidnapping and burglary are common, don't accept a room that has a balcony or is low to the ground (e.g., on the first or second floor). Look out the windows of your room to see if they can be accessed easily via the top of a wall or fence or from a fire escape. In high-crime cities, don't stay in a room looking out on dimly lit areas. If the chances are high that a terrorist bomb could explode in front of the hotel, don't accept a room that faces the street. If arson is a worry, don't stay above the fifth floor, because most fire ladders don't extend much higher. And make sure all doors and windows lock properly.

 Hotel rooms on the third to fifth floors are preferred for maximum safety and ease of fire rescue or emergency escape.

 Don't mention your room number when in places where you might be overheard, such as in elevators or restaurants.

Reject hotel rooms with sliding doors.

Rooms with sliding doors can make for great views or nice strolls outside. But, frankly, how often do you really use them? Once, maybe. Sliding doors are open invitations to criminals, kidnappers, and terrorists. Put your safety first, and get another room.

✔ If you have no choice but to stay in a room with a sliding door, not only lock it and use the security bar, but also use an empty dresser drawer for added safety. Place the drawer lengthwise on top of the door track and push it squarely into the corner of the doorframe. That way, if someone forces the door open from the outside, he won't be able to get it open completely; the dresser drawer will be in the way. Don't use broom handles or small pieces of wood to block sliding doors. An expert burglar will lift those out with a coat hanger.

After getting your key, always be escorted to your room by a hotel employee.

Once you've checked in and gotten your key, insist on getting someone to escort you to your room and ask him to wait while you do a bit of reconnaissance. Check every place a criminal or terrorist might be lurking in wait. Look in the bathroom, closets, behind drapes, and under the bed. If there are connecting doors to adjoining rooms, make sure they're locked. Only then dismiss the bellhop (with a tip). It would be prudent to follow the same procedure each time you return to your room, keeping the door ajar until you're satisfied no one else is there.

Make your first outing a walk to the nearest fire exit.

On the back of most hotel doors is a floor diagram showing your room location and all emergency exits. Grab your key and take a stroll to the nearest emergency exist, being sure to count the number of doors you pass along the way. Keep a mental note of the route. And if there is a fire and the halls fill with smoke, crawl along the floor, counting the doors, until you reach the exit. Never use an elevator in a fire, because the doors could open on the fire floor and you'll be burned to a crisp.

Know what to do in a fire.

In a hotel fire, it's vital to remember to do one thing first: Press your hand against the top of the door, the doorknob, and the crack around the door to feel for heat. If the door is hot to the touch, *don't* open it. There's most likely a ball of fire on the outside that could consume you in an instant. Fire can travel at upward of 19 feet per second. If the door is hot to the touch, call the desk for help. Wet some towels or pillowcases and stuff them into door cracks to keep smoke out of your room. Swinging a wet towel around the room can catch smoke particles.

Look out the window to see if any rescuers have arrived. But don't open your window all the way unless the smoke becomes unbearable; an open window can create a natural draft that could suck smoke into your room. Open a window from the top to let out heat and smoke and at the bottom for fresh air to breathe. Cover your mouth and nose with a moist cloth and wait for help to arrive. Signal emergency personnel from your window to show them your location. Wave a sheet out the window. Use the hotel phone or your cell phone to help firefighters find you; you could flash a mirror or a lamp to signal your whereabouts to rescuers. If you have a whistle, use it. For more information on fire safety, go to the U.S. Fire Administration's Web site at www.usfa.fema.gov.

 If you must exit through smoke, clean air will be several inches off the floor. Get down on your hands and knees, and crawl to the nearest safe exit.

🚫 Don't go back into a fire for anything! Your life is your most valuable possession.

Stow your shoes, robe, wallet, and passport next to your bed.

In an emergency, you may have to dash out of your room in a hurry. The less time that takes, the better. So keep everything you *really* need near to hand. Keep your wallet and keys (and passport, when overseas) on the nightstand next to your bed. Put your shoes and a robe within easy reach. You may also want to keep your briefcase and laptop similarly nearby. But whatever you do, don't try to take all your belongings with you in an emergency evacuation. Things can always be replaced. You can't be.

✔ Store your wallet and other important items in your room safe while bathing. If you're abroad and want to conceal your U.S. citizenship from causal observers, buy a plain cover (available at most stationery stores) for your passport.

Keep your belongings tidy so you can leave a hotel at a moment's notice.

You never know when you may be called away suddenly or need to leave your hotel in a hurry. When time is of the essence, you don't want to be scurrying about your room, gathering up scattered clothing, papers, and toiletries. So keep things neat. And if you've left clothes to be cleaned, don't worry about them; they can always be shipped to you later. Depending on the situation and reason for your early departure, you may or may not want to leave a forwarding number and address with the front desk. If you suspect that the information could fall into the wrong hands, simply tell the desk to e-mail you or call your office should the need arise.

🚫 Be sure to inform your family, your company, and the people you plan to meet (as well as the local U.S. embassy or consulate, if you're overseas) of your move. You don't want them worrying if they call the hotel and find you're no longer there.

Inspect the contents of your luggage before checking out of your hotel.

One way terrorists could plant a bomb aboard a plane would be to tamper with an unsuspecting traveler's luggage while it was still in his hotel room. So inspect the contents of your luggage before checking out of your hotel. If you find anything suspicious, leave the room immediately and then call the police.

Be suspicious of unexpected mail.

If you receive unexpected mail at your hotel or branch office, treat it with suspicion. Follow the mail-handling procedures laid out in the earlier chapter on personal safety. Recognize, too, that mail from an unknown source means that someone knows you're in town. This could pose a risk of abduction or worse, especially when you're overseas. If you're worried, contact the police or the local U.S. embassy or consulate, and consider changing hotels—and don't leave a forwarding address or phone number other than that of your home office.

 Be wary, too, of any packages left outside your hotel room and envelopes stuck under your door or left on your car windshield.

Be careful about hanging signs on your hotel door.

Normally, signs like Do Not Disturb or Make Up the Room are placed on hotel room doors to provide instructions to the staff. But in this age, they also provide valuable information to would-be thieves and attackers. Forgo placing any signs on your hotel room door. Call to request service; that goes for meals as well. Don't leave a breakfast list out; call instead.

When answering your hotel room door, always keep the door chained or blocked with a doorstop.

You may have ordered room service or be expecting a guest, but how can you be sure that a criminal doesn't lurk on the outside of your hotel door? A look through the peephole may be little help if you can't see the person on the other side or if you don't know what the person is supposed to look like. So the best thing to do is to keep the door between you and a visitor for as much time as needed to determine that he doesn't intend you any harm. There are two easy ways to do this: One is to always keep your hotel room door chained. To greet a waiter or a guest, open the door first with the chain on to ascertain that the person is, in fact, whom you expect him to be—and not a criminal expecting to push his way into your room. A second way is to carry a rubber doorstop with you. Although most hotels have chains on their doors, you may want added protection. By prepositioning a doorstop within inches of your door, any attempt at forcible entry will be stymied long enough for you to cry out for help and get to the phone to call hotel security. Rubber doorstops are cheap and light and can found in most hardware stores. Don't hesitate to buy one and make it part of your travel kit. Alternatively, portable travel locks can be used to secure hotel doors.

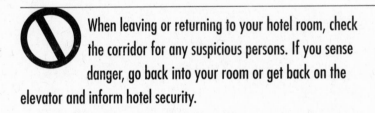

When leaving or returning to your hotel room, check the corridor for any suspicious persons. If you sense danger, go back into your room or get back on the elevator and inform hotel security.

Be especially alert on stairs and in elevators in unfamiliar buildings.

While on a trip, you'll probably enter many buildings for the first time. This means that your knowledge of building security will be extremely limited. You won't necessarily know, for instance, if a building has a history of robberies or muggings. So be on guard when moving about unfamiliar buildings—and that includes your hotel. Of all the places you might be attacked, a building's stairs and elevators are the most likely spots. Before taking the stairs, look up and down the staircase and listen for any unusual noises. If you suddenly hear footsteps moving in a hurry, exit at the nearest floor and wait until the commotion stops. If you spot someone lurking below, again exit and wait until the coast is clear. And never use a staircase where the doors automatically lock behind you. When an elevator's doors open, don't get on if you see someone or something that doesn't look right; wait for the next car. If you're already in an elevator and a person enters whose look disturbs you, get off immediately.

✓ When you're in an elevator, especially in an unfamiliar building or in a high-risk city, always position yourself within reach of the emergency alarm.

Never walk down dark alleys or empty streets.

Dark passageways and deserted streets are notorious for abductions, robberies, muggings, and sexual assaults. The criminals know the territory and position themselves to take a captive, rob a passerby, or rape a woman. Their tactic is to use surprise to gain the advantage. So do your homework. Study quality maps showing the main thoroughfares near your hotel or branch office. Always walk down well-lighted streets, and if possible, use routes with a lot of other pedestrians or plenty of cars passing by. Kidnappers tend to avoid well-lighted, heavily traveled streets, knowing there are better spots in town to victimize the unwary. Ask at your hotel or branch office about the parts of town you should avoid. You don't want to get off a subway, say, and suddenly find yourself in the midst of thieves.

✓ Carry adequate travel insurance, covering theft, lost luggage, and the like; comprehensive annual policies are the most cost-effective for frequent travelers.

Ask at your hotel for the names and numbers of at least two car services.

It's easy to find yourself in a place in a strange city where there simply aren't any taxis. It may be that the hour's late or you may have strayed off the beaten track. So make it a point before leaving your hotel to ask the desk for the names and telephone numbers of at least two reliable car services. Besides offering convenience, those numbers could prove to be lifesavers if you end up in a crime-ridden part of town, particularly late at night.

✔ Keep the address and phone number of your hotel with you at all times. This is particularly important overseas, where getting a phone number from directory assistance can be a challenge if you don't speak the language.

CARJACKING SURVIVAL

Every year, at least 49,000 carjackings are attempted or succeed in the United States, according to the latest U.S. Department of Justice data.

Spot a carjacking before it happens.

Your times of greatest vulnerability to a carjacking are when you're slowing down or stopped, say, at a light. It's imperative that you keep all car doors locked and all windows rolled up in high-risk cities and foreign countries. Carjackers will often put up an artificial roadblock. It could be debris in the road, a felled tree, or a car blocking passage. If you encounter any of these, don't get out of your car to inspect the problem, and always leave enough room in front of you—preferably, two car lengths—so you can turn your car around. If a car is ahead of you, an easy way to gauge the correct distance is to stop far enough behind so you can still see where the car's back tires meet the road. Many times carjackers will have an accomplice pin you in from behind to prevent you from backing out of the ambush, so again leave as much leeway in front of you as possible.

Be on the alert for strangers lurking about a cash machine. Carjackers stake out automatic teller machines (ATMs), particularly those located in dark, isolated areas. If you need cash at night, find an ATM in a well-lighted area, preferably near an open business establishment. Other venues for carjackings are public garages, mass-transit parking lots, shopping malls, 24-hour grocery stores, self-serve gas stations and car washes, and highway exit and entrance ramps.

> ✔ If you spot suspicious activity at the intersection ahead, slow down and try to time the light, so you can drive straight through without stopping.

> 🚫 Beware of bump-and-run carjackers. These bandits bump you from behind, and when you go to inspect the damage, an accomplice gets in your car and drives off.

Frustrate a carjacker's plan by throwing your car keys away.

If you have left your car unlocked and a carjacker gets in, immediately turn off the ignition and throw the keys as far as you can, either through an open window or by opening the door. By throwing the keys away, you'll have defeated the carjacker's plan and knocked him off-balance. Now he can't force you to chauffeur him (which was probably his plan). Without keys to the car, most likely he'll abandon the effort and flee.

> 🚫 Don't be the first to run out of the car, because the carjacker may shoot you in the back. Wait until he starts to get out of the car, then you do the same. Run away from the car and seek help. There's no shame in running away. It's the smart thing to do for your own protection.

If faced with a weapon, comply with the carjacker's demands.

In a life-or-death situation, accede to the carjacker's demands. Don't argue. Preferably, you'll have had time to get the keys out of the ignition. In which case, hand them over and ask if you can leave. If you simply try to run away without first asking permission, the carjacker might shoot you. If he doesn't respond to your request, try to win his sympathy by telling him you have a family.

✔ **If you're driving and the carjacker fires a gun at police, intentionally crash your car. You're more likely to survive the impact of an elective crash than an exchange of gunfire.**

🚫 **Never make any promises to your attacker. Don't say, "I promise I won't tell the police what you look like," because that will make him more nervous and fearful, and he might kill you to eliminate you as a witness.**

As for corporate travel procedures, companies of all sizes should establish new safety and security procedures to ensure that their executives and employees are safe while traveling on business and provide a ready means of communication in emergencies.

Require business travelers to submit detailed itineraries.

Businesses should know where their traveling employees are at all times. Companies of all sizes should require employees to submit detailed travel itineraries prior to departure, primarily to help to ensure the traveler's safety and to assist law enforcement if the employee goes missing. Requisite information should include flight schedules, hotels, rental cars, and the names, addresses, phone numbers, times, and dates of all scheduled meetings. Travelers should update this information, via phone or e-mail, as circumstances warrant.

✓ Self-employed business travelers and vacationers should establish a similar routine, leaving an itinerary and contact information with a relative or friend.

Every firm should have a dedicated travel manager.

Large corporations, of course, have whole departments that handle business travel, but many small firms don't. Given the uncertainties and risks associated with terrorism, every company should have a dedicated travel manager—even if the task is assigned to someone as a part-time adjunct duty. A travel manager serves a multifunctional role. In some cases, he'll book flights and make hotel and rental-car reservations. But his most important responsibilities should be to stay abreast of the latest terrorist threats in the United States and around the world, to provide updates to traveling employees periodically, to know where traveling employees are at all times, and to serve as *the* contact person in emergencies.

Create 24-hour hot lines and recorded threat updates.

Business travelers on the road need to feel connected to their home offices. Three good means of accomplishing this are (1) create a 24-hour hot line that employees can use to reach someone at the company. Establish a backup telephone number as well in case the local telephone system fails; (2) provide prerecorded voice messages on the latest terrorist threats, both in the United States and overseas, that traveling employees can call. The State Department, Federal Aviation Administration, FBI, newspapers, television, and newswires should serve the sources of the information; (3) offer the same hot line and information services via the Internet, using e-mail and password-protected Web pages.

Rethink security procedures involving your corporate aircraft.

Many executives are avoiding commercial aviation in favor of corporate aircraft to forgo the delays associated with post-9/11 airline travel. These same executives, however, may unwittingly be exposing themselves to other dangers, such as tampering with parked corporate jets, hijacking, or abduction for ransom. Corporations with their own aircraft need to rethink security. All luggage and parcels should be identified before being put on a plane. Especially when overseas, never rely on airport security to guard your parked jet. Hire your own security force to stand watch. And make sure that your caterer is reliable. In a foreign country, it may be best for a member of the crew to pick up food and beverages at your hotel for consumption in-flight.

Take out ransom insurance and have security-service firms on call.

The danger of kidnapping for ransom, particularly in countries at high risk of terrorism or violent crime, means that most every firm should have ransom insurance. In addition, companies should maintain a list of professional security firms in the United States and overseas that could supply bodyguards or otherwise assist a traveling employee who's in danger or in distress.

KIDNAPPING SURVIVAL

Kidnapping is much more common abroad than in the United States. In recent years, Colombia, in particular, has become notorious for kidnappings for ransom. In Russia, too, several American business travelers have been kidnapped and even murdered in recent years.

Here, according to the U.S. State Department, are the countries in which kidnappings occur most frequently: in Africa, Angola, Democratic Republic of the Congo, Ethiopia, Kenya, Nigeria, Rwanda, Somalia, and Uganda; in Asia, Georgia, Indonesia, the Philippines, and Yemen; in Europe, Russia; and in Latin America, Colombia, Ecuador, El Salvador, Honduras, Mexico, Nicaragua, and Venezuela. For country-specific information on kidnapping, consult the U.S. State Department's travel advisories online at http://travel.state.gov/travel_warnings.html.

Know when you're being targeted for abduction.

You may be able to spot telltale signs that you're being targeted for abduction. These include people who are clearly taking too great an interest in you; repeated sightings of the same people observing you, say, while entering and leaving your hotel or branch office; strangers who have asked others about you; an accidental encounter with a stranger who then tries to strike up a conversation, asking who you are, where you come from, what you do for a living, and how much money you make; and any sign that you are being followed on foot or trailed by a car. Immediately inform the U.S. embassy, police, your company, and your hotel of your suspicions. You might consider cutting your trip short and returning at a later date to complete your business. And remember, too, that a kidnapping can take place anywhere—in your hotel room, on the street, and even in a taxi.

 Don't invite extortion by engaging in activities that are illegal or compromising.

Try to attract attention if you're being kidnapped.

You may get only one chance to prevent a kidnapping, so try to attract as much attention as possible if the attack takes place in public. If you're in your hotel, yell. And yell, "Fire!" Other guests are more likely to respond to a scream of "Fire!" than to a cry of "Help!" However, if faced with a gun or knife and the risk of injury or death is great, go quietly.

Conserve your strength and look for landmarks if you're bundled off.

Know that if the kidnapper succeeds, you may be blindfolded, knocked unconscious, or drugged. You might even be forced into the trunk of a car. If you're conscious, don't struggle; conserve your strength, for it's likely your meals in captivity will be few and far between. If you're in a vehicle, see if you can spot any landmarks along the way to help authorities later.

Don't look your kidnapper in the eye.

Never look your kidnaper in the eye. That's the first rule to follow in a kidnapping or any hostage situation. Direct eye contact intimidates an abductor and creates in him a fierce animosity toward you. He may decide to treat you particularly harshly or violently. He could bind you painfully tight and threaten you with death. Instead, be cooperative. Don't ask questions. Don't get angry. Don't be antagonistic. Rather, be controlled and reserved. Tell yourself to be calm and that everything will work out all right. Pray, if you're so inclined.

Be humble and try to gain the sympathy of your kidnappers.

If kidnappers are holding you for ransom, be humble and appear helpless. Don't do anything to make your captors nervous or afraid. Don't act like you're James Bond. Don't threaten them with retaliation and revenge, because they may decide to kill you. Tell your abductors, "I'm not a hero, and I'm not trying to be a hero." Never demean or insult them, and never invent stories about being wealthy and important or offer to pay them off. Don't put on airs and say things like "I'm very rich. I have jewelry at the hotel. Take me there and I'll give it to you." If you make yourself out to be rich, your abductors will never believe you if you admit to them later that you really are poor. Abductors do make mistakes and occasionally kidnap someone who has little or no money to meet their ransom demands. If you find yourself in such a situation, first let things play out a little. Your kidnappers will need time to relax before you can attempt to reason with them. Then, tell them the truth: "I want you to know, contrary to what you think, I'm not rich. I've got a wife and kids like you. I'm just here trying to make a living."

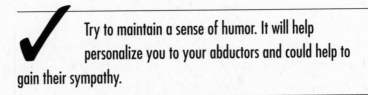

Try to maintain a sense of humor. It will help personalize you to your abductors and could help to gain their sympathy.

Assess your chances of escape.

Time and again in abductions, a kidnapper will let down his guard as the ordeal wears on. This could present you with a chance to escape or overpower him. If you mean to flee, see if your abductor is preoccupied and not looking in your direction. Determine whether you can release yourself if you're tied up. Look for a ready avenue of escape. Be certain, however, that you're physically and mentally attuned to the task. If you have doubts, it's better to do nothing. In general, we don't recommend attempting to overpower an assailant, unless you are clearly up to it. It's vital that you be physically fit and trained, preferably in the martial arts, if you intend to take down your captor.

If you mean to escape, be certain that you're in a good position to carry it off. You must time an escape well to have any chance of success. Begin by gaining the kidnapper's trust. Choose your best route of egress, say a window or door, and move nearer and nearer to it on successive occasions over days or even weeks. The kidnapper will grow used to your behavior. Then, when the time's right—say when he's far away from you and distracted—go for it. Be careful, however. A mistake on your part could cost you your life. You have to assess the situation thoroughly and determine whether you can be effective in your action, be it trying to overpower your captor or running away. If you've determined that your attacker is a professional, don't do anything. Just sit there. A professional is prepared for attempts by his captives to retaliate or escape. Even long into an abduction, his attention span, emotional balance, and physical strength haven't been sapped. A well-trained terrorist or professional kidnapper simply isn't going to make the kinds of mistakes that would give you an opening to either disarm him or escape.

If you manage to escape, don't try to hide.

It's natural to hide from danger. But that would be a mistake if you just escaped a kidnapper. If you run and hide, say in the woods, the first person likely to find you is your abductor. So rather than hide, you must do the precise opposite: run toward people, houses, populated areas, or business districts. Stay out in the open where people can see you. If you're in a remote, wooded area, look for roads, listen for the sound of traffic or trains, and scan the horizon for buildings or spires.

While a hostage, try to stay mentally and physically fit.

Your mental and physical health is bound to suffer in captivity. You'll probably be held in a small, confined space and may be tied up or chained for long stretches. You'll be fed infrequently, the quality of the food will be inferior, and the portions will be small. Kidnappers do this for a reason: to weaken you. It's imperative, therefore, that you try to stay as fit as possible, both mentally and physically. Exercise if you can, and use your mind. Don't dwell on your plight. And remain hopeful.

If you receive a ransom demand, don't use the same phone to call the police.

Kidnappers making ransom demands check to see if the police are being called. They do this by waiting a few minutes after making their initial call and then calling the same telephone number back. If the phone line is busy, they assume you're talking with the police. Should you receive a ransom demand, call for help from another phone—a cell phone or a neighbor's phone. But do call the police and the FBI. Never try to handle a demand for ransom on your own.

✔ Contact the employer of the person who's being held for ransom. Many companies have ransom insurance. Interestingly, many ransom insurance policies require policyholders never to announce that they have such insurance or to say which employees are covered by the plan. It could turn out that the person being held hostage is covered by ransom insurance but doesn't know it.

Take out ransom insurance if your employees travel overseas often.

It makes no sense to be penny wise and pound foolish in this age of terrorism and kidnapping for ransom. Consider the plight of Thomas Hargrove, a

Texas science writer. He was working in Colombia when he was kidnapped in September 1994 by the Revolutionary Armed Forces of Colombia. He was held hostage for 11 months, until his family reportedly came up with a $500,000 ransom payment. Ransom insurance is widely available. (The names of several insurers are listed in the appendix of this book.) Companies that frequently send employees overseas or have foreign subsidiaries are the ones most in need of ransom insurance.

Know when to hire professional bodyguards.

In November 1986, the president of France's Renault automobile company, Georges Besse, was shot to death in Paris by the French Marxist-Leninist group Action Directe. Three years later, Deutsche Bank chairman Alfred Herrhausen was assassinated in Frankfurt, Germany, by terrorists with the revolutionary Red Army Faction. The point is, if you're a target and you don't know it, the terrorists have got you. So you have to know when to get professional help. The risk of being attacked is greatest for executives of companies viewed as symbols of the United States or any country that's an integral part of the global coalition against terrorism. Terrorists are publicity seekers. The bigger the name of your company, the higher your position in that company, and the better known you are, the greater risk you face. It may be that you only require advice on how to protect yourself and what to avoid overseas. Others may need full-time bodyguards.

TIPS FOR WOMEN

Women represent a growing percentage of business travelers. Furthermore, many of today's business assignments take women to unfamiliar and sometimes dangerous parts of the world. Then, too, when traveling on vacation, more and more women are opting to make unescorted sojourns.

What concerns female travelers most is the sense that criminals view them as easy prey. This is indeed a serious problem. If you feel you're likely to become a victim of crime, you aren't going to function well as a business traveler—particularly when it comes to traveling abroad—and you're going to be reluctant to travel alone (even within your own country). The best way women can overcome any fear or trepidation about traveling—whether it's for business or pleasure, at home or abroad—is to take the proper precautions. The suggestions listed below are the best techniques we know of for women to stay safe when traveling—especially, for women traveling alone.

Our recommendations serve a dual purpose: not only are they practical tips that can help to protect you against crime, but they'll also serve to boost your self-confidence. Simply knowing that you've taken every possible precaution lowers travel anxiety. And a person who is self-confident is less likely to become a mark for a criminal. Criminals are good at spotting people, particularly out-of-towners or foreign visitors, who show signs of fear, confusion, overanxiety, or bewilderment. They see any of these signs in a traveler's demeanor—particularly, a female traveler's demeanor—as increasing their chances of getting away with a crime. In a sense, fear of becoming a crime victim may actually increase the likelihood that you will become a victim. So the best thing you can do for yourself is to lessen your uncertainty and anxiety, especially about traveling to places that you've never been to before or to parts of the globe known to be high-risk destinations.

The following steps aim to help you lower your anxiety and boost your confidence by teaching you how to stay safe when away from home. The techniques are tested and are known to defeat criminals. Making them a habit will lower the chances that you will fall prey to a criminal and will also give you more peace of mind about traveling on business or for pleasure. We further recommend *Safety and Security for Women Who Travel* by Sheila Swan and Peter Laufer, which discusses the things you should do before you leave and en route to foreign destinations, as well as how to cope with difficulties after you've arrived at your destination. The paperback, published in 1998 by Traveler's Tales, doesn't address post-9/11 issues. However, it remains a useful guide for women travelers. Excerpts from the book are available online at the Traveler's Tales Web site (www.travelerstales.com).

Use only your last name and first initial when booking a hotel room, and don't let the world know your room number.

Criminals are always on the lookout for women traveling alone. One way to reduce your exposure is to book your hotel using only your last name and first initial. Also, don't make any other reference to your gender, such as Ms.,

Miss, or Mrs. When you're handed your room key, if the desk clerk announces your room number loudly enough for an eavesdropper to hear, ask for a different room and tell him not to make the same mistake twice.

If crime is bad at your destination, conduct business meetings at your hotel—and book a suite!

In some cities and foreign countries, it simply doesn't pay to travel around much. Crime can be so rampant that no matter what precautions you take, you're likely to become a crime victim sooner or later. In such cases, the wisest tack is to turn you hotel space into a quasi office and conduct your business affairs there. That way, you'll have the security of the hotel to protect you. But book a suite. You don't want guests getting the wrong idea when you invite them to your room, and keep the bedroom door closed to reinforce the point. Most important, only let guests into your room that you absolutely trust; if you have any doubts, conduct your meeting elsewhere in the hotel. You might find a quiet place in the hotel lobby to discuss business, or you could invite your guest to a business meal at the hotel restaurant. Be mindful of local customs and attitudes, however. In some places around the world, inviting a man to dinner is tantamount to a proposal of marriage—and sometimes marriage without the ring.

✔ We recommend that you stay at a U.S. hotel chain when overseas. An American hotel is likely to have better security (and perhaps better rooms) than the average hotel overseas.

Avoid walking the streets alone after dark and always stride with confidence—and a whistle.

Nighttime is not the right time to be out and about, unescorted, particularly in a foreign city where the threat of crime or terrorism is significant. If you must be out in the dark, take taxis or call a car service. But no matter the time of day, whenever you're on the streets, always walk with sense of purpose, displaying self-assuredness and resolve. Your body language will make you appear less vulnerable, and a would-be assailant may well let you pass and wait for easier prey.

It helps, too, to know where you're going. So make a daily itinerary, map out your routes beforehand, and remain confident along your route. If you start to feel anxious about your surroundings, walk close to other people. That doesn't mean you have to strike up a conversation. Merely stick close enough to make it appear that you are part of a larger group. And, if you do need to ask a question of a stranger, approach a woman. She's more likely to be sympathetic—and protective—than a man.

You should also consider carrying a whistle. Attackers hate noise and may well flee if you sound an alarm. Be sure, however, to keep your whistle someplace handy. It defeats the purpose if you have to fish for it in the bottom of your handbag. Pepper spray is another option, but not one that we recommend. Your aim is to get away from any assailant, not engage him in combat. Also, when overseas, laws may ban such devices.

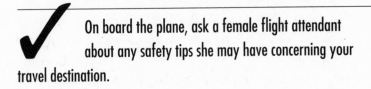

On board the plane, ask a female flight attendant about any safety tips she may have concerning your travel destination.

Never keep important documents or large amounts of money in your purse.

Purses are the first things that thieves go for. Many thieves don't even need to use force or the threat of violence to separate you from your handbag. They'll simply snip your purse strap and walk away with your bag. Many are so good at purse snatching that you won't notice what's happened until it's too late. Still, it helps to have a bag with a strap that's long enough for you to wear across your neck and chest. While you may not be able to prevent a skilled thief from absconding with your purse, you can at least limit the damage by not keeping important documents (including your passport), large sums of cash, and all of your credit cards in your purse. Vital documents might be kept in a briefcase or any travel bag that you have to hold in your hand. It's much harder for a thief to wrestle away something that you have a grip on. Also, try to wear a piece of clothing that has pockets (preferably on the inside), or buy a money belt or waist pack to hold your passport, extra cash, and at least some of your credit cards. Money belts can be found in stores that cater to travelers and can also be bought online at such sites as Amazon.com.

> ✔ When walking down streets, keep your handbag on the side of your body that's away from the road to avoid someone in a passing car snatching it; on escalators, keep your bag away from the escalator that's moving in the opposite direction.

Wear sunglasses to avoid unwanted attention.

One of the more interesting tips for women travelers that we've heard is to wear sunglasses so as to avoid unwanted advances. Sunglasses, of course, give an air of self-confidence. They are a way of appearing above the fray. More important, sunglasses hide your eyes. Anyone seeking to get your attention needs to make eye contact. Without that, you can simply walk away and the chances are the interloper won't follow.

Think twice before taking orders from a policeman, particularly in a foreign country.

In their book, *Safety and Security for Women Who Travel*, Swan and Laufer urge women to take a moment to assess any order from a policeman or other authority figure. "Remember," they say, "some will be using their badge of office to try to take advantage of you and will consider you especially vulnerable because you are a foreigner and a woman." In developing nations, they add, "Rogue policemen may offer to drop serious charges against you for sexual favors." Refuse the offer, "politely but firmly," and take your chances with the criminal justice system, Swan and Laufer advise—and we concur.

We would add that you should not allow yourself to be bullied by a badge. Be courteous and respectful, but not pliant or afraid in your dealings with police. If you're overseas and the situation warrants, tell the officer you intend to call the U.S. embassy and point to your cell phone. But *never* grab for the phone before telling the officer what you intend to do. He might think you're reaching for a gun.

✔ **If at night police signal for you to pull your car over, turn on your directionals and drive slowly to a well-lighted area before stopping.**

Dress and act appropriately when overseas to avoid encounters with police.

Women travelers abroad should act and dress "conservatively," Swan and Laufer say, because "your appearance can instigate unnecessary interactions with authorities." In Muslim countries, for instance, loose-fitting clothes that cover your arms and legs would be appropriate. The authors further note that women are not welcome everywhere overseas; some places, notably religious shrines, specifically enjoin women from entering. So follow local customs if you want to avoid frivolous encounters with the police. Swan and Laufer also offer another helpful tip: spend a little time at a local café and observe how the local women dress and behave, particularly with men. That is an excellent way to gain insight into a culture.

Take precautions to reduce your risk of assault.

MIND-SET Protecting yourself begins with keen awareness. Always be alert wherever you go, and use your head. Walk with confidence and a sense of purpose. Let no one get the impression that you're wandering around aimlessly and alone. Maintain a constant awareness of your surroundings. Look around to see who's near you and what's going on in your vicinity. Go with your instincts. If a situation or location makes you uncomfortable or uneasy, don't think twice. Leave! Rationalization can get you in trouble.

Don't let alcohol or drugs cloud your judgment. If you are at a bar or reception, always keep your drink in your hand. If you step away and return to your table or seat, ask for a new glass of whatever it was you were drinking. Reserve your trust for only a few people. You never know who may have slipped something into your drink while you were away. And that goes for people who work at your company, as well as clients and strangers.

HOTEL ROOM Upon arriving in your room for the first time and then every time you return, keep the door open until you've checked the premises for anyone who may be hiding in wait. You can prop the door open with a piece of luggage, always left nearby, or a doorstop. Avoid using your handbag or briefcase for this purpose, as someone might snatch it from behind your back. Also, when you first arrive in your hotel room, make sure that the locks work on all widows and any doors to your room, and whenever you return, check to see if the locks are still in the same position as when you left.

After you're satisfied you're alone, chain the door behind you and use a rubber doorstop you've brought along as backup. When answering the door, keep the chain on until you know for sure who's on the other side. Never open your door to strangers. When in doubt, don't be embarrassed to phone the front desk for verification. Don't walk about the hotel corridors looking for a soft-drink machine or ice dispenser; call room service. Keep a light on in your hotel room at night—even when you're sleeping. Don't say, "I can't sleep with the light on." Practice the technique at home; you'll get used to it.

WALKING Avoid walking or jogging alone, especially at night. Stay in well-traveled, well-lighted areas. Wear clothes and shoes that give you freedom of movement. Be careful if a stranger asks for directions. Keep your distance from the person or the car he's in. Walk down the middle of sidewalks, staying clear of hedges, doorways, and alleyways—and passing cars. And walk facing traffic. That makes it impossible for someone to follow right next to you in a car. It also makes it harder for someone to pull you into a vehicle. If you think you're being followed, change direction and head for an open store, restaurant, or even a lighted house. If you see a police officer, tell him of your concern. You could also hop aboard a bus or hail a cab. Otherwise, approach someone, preferably a woman (or a couple), and say that you feel a bit uncomfortable walking alone and ask if she'd mind some company. It's probably not wise to say that someone's following you, because she, too, may become afraid and avoid you like the plague.

DRIVING   If you've rented a car, have your key ready before you reach the car. Look around the car as you approach. Before getting in, check the backseat for anyone who might be lurking there. In fact, before you even leave your car, move up the front seat, so you can see more easily into the backseat area when you return. Also, check to see if the front seat has been returned to its original position. If it has, leave immediately and get help.

If your car is in an underground garage, ask the front desk for an escort. Keep your car facing outward in parking lots to make getaways faster and easier, and park as close to an exit as possible. Always park in well-lighted areas, and keep car doors locked at all times. On the road, never roll down any window other than your own, and even then, never roll down your window all the way. Don't allow for enough room for someone on the outside to get his arm in the car. Never pick up hitchhikers. And, of course, don't hitchhike yourself. If your car breaks down, call the rental company for assistance. Be wary of offers of help from strangers. Don't stay in your car if you don't have to; go to a nearby shop or café—and tell the rental company where you'll be. If assistance is unlikely to arrive quickly, get a cab or car service to take you back to your hotel. If you've broken down on a highway or a deserted road, call for assistance and stay put until help arrives, keeping your doors locked. If you don't have a cell phone or if you can't reach the rental company, lift up the hood, get back in the car, lock the doors, and turn on your flashers. If an emergency call box is nearby, use it.

✔   If you need to get someone other than your rental company to fix the car, rental firms will often reimburse your expenses, so ask for a receipt from the repairman.

IF ALL GOES WRONG

The State Department offers a fair amount of information and help to U.S. citizens (and their relatives) who find themselves in the midst of a crisis while traveling abroad. Be sure to take advantage of it.

Heed official U.S. warnings to defer travel to or to leave a foreign country.

It doesn't happen often, but the U.S. State Department does on occasion instruct American citizens to avoid traveling to or staying in a country that is in violent turmoil or on the brink of war or where the lives of Americans are directly threatened. In June 2002, for example, as tensions between India and Pakistan escalated, the State Department issued travel warnings for Americans to "defer all but essential travel" to those countries and authorized all nonessential U.S. personnel and their families to leave. And after a June 14 car bombing in Karachi, State further said that it "strongly urge[d]" American citizens in Pakistan to depart. Prudence dictates following such official advice. The State Department's travel warnings are posted at http://travel.state.gov/travel_warnings.html.

Map at least two emergency escape routes out of a foreign country.

Nowadays, you can never tell when a country may erupt in violence or become the target of a concerted terrorist campaign. You may, therefore, need to leave a foreign country in a hurry. Plan an exit strategy, based on at least two different routes of escape. Designate where you'd go and how you'd get there in case of an emergency. If the situation is extreme, you may want to eschew commercial air travel, for fear airports and air travel will be dangerous. If rioting breaks out and if Americans are singled out, airports will be not only chaotic but also prime targets for attack. Your best bet may be to travel by road to a neighboring country and hop on a plane from there. This will require some preplanning. If you have a rental car, get a map and determine a route to the safest neighboring country. If not, think of someone who might drive you (e.g., a company employee residing in-country, a client, a car service, or an off-duty hotel employee). And prepare a small survival kit filled with essentials (e.g., medicines, extra glasses, food, and water) that you can grab in a hurry. Contact the local U.S. embassy or consulate for advice; you might also ask if you could get a ride out of the county with a member of the diplomatic staff. Also, see the State Department's advice on evacuations at http://travel.state.gov/crisismg.html.

✔ **If you have no other choice but to fly out commercially, pick a destination other than the United States. Select a safe-haven country, like Switzerland. Note, though, that some of these countries require visas. So check in advance about any entry requirements or restrictions.**

Get help from the State Department if you're stranded overseas and need financial assistance.

If you're a U.S. citizen and run out of money while overseas, the State Department will lend you a helping hand. Destitute or stranded Americans should contact the Overseas Citizens Services of the Office of American Citizen Services and Crisis Management at 202-647-5225 or through the local U.S. embassy or consulate. More information is available at http://travel.state.gov/finance_assist.html.

Contact the State Department if a relative becomes embroiled in a foreign crisis.

Families whose U.S.-citizen relatives are directly affected by a foreign crisis should contact the Department of State through its Office of American Citizen Services and Crisis Management at 202-647-5225. If a 24-hour task force or working group is established in the Department of State Operations Center to manage the crisis, you'll be directed to the Task Force at 202-647-0900.

For more information, see "Crisis Awareness and Preparedness" at http://travel.state.gov/crisismg.html.

Should a death occur overseas, be prepared for a complicated and costly process to get the remains returned to the United States.

The State Department's Office of American Citizen Services and Crisis Management (ACS) says that approximately 6,000 Americans die outside of the United States each year and that the majority of these are long-term residents of a foreign country. ACS assists with the return of remains for about 2,000 Americans annually. (Call 202-647-5225.) When an American dies abroad, a consular officer notifies the next of kin about options and costs for the disposition of the remains. Costs for preparing and returning a body to the United States are high and are the responsibility of the family. Often local laws and procedures make returning a body to the United States for burial a lengthy process.

Don't expect too much official help if you get arrested abroad.

If you're arrested in a foreign country, you're pretty much on your own. The ACS says that more than 2,500 Americans are arrested abroad annually. More than 30 percent of these arrests are drug-related, and over 70 percent of drug-related arrests involve marijuana or cocaine. The rights an American enjoys in the United States "do not travel abroad," notes the ACS. "Each country is sovereign and its laws apply to everyone who enters regardless of nationality. The U.S. government cannot get Americans released from foreign jails" ("When You Need Help," Office of Overseas Citizens Services, Department of State Publication 10252, revised May 10, 2002). A U.S. consul will insist on prompt access to an arrested American, provide a list of attorneys and information on the host country's legal system, offer to contact the arrested person's family or friends, visit regularly, protest mistreatment, monitor jail conditions, provide dietary supplements, if needed, and keep the State Department informed.

ACS is also the point of contact in the United States for family members and others who are concerned about a U.S. citizen arrested abroad. A toll-free hot line at 888-407-4747 is available from 8 a.m. to 8 p.m. Eastern time,

Monday through Friday, except U.S. federal holidays. Callers who are unable to use toll-free numbers, such as those calling from overseas, may obtain information and assistance during these hours by calling 317-472-2328. For after-hours emergencies, Sundays, and holidays, call 202-647-4000 and request the Overseas Citizens Services (OCS) duty officer.

✔ **Relatives, friends, or colleagues can use the same phone numbers to try to track down a U.S. citizen who has gone missing overseas. The ACS fields about 12,000 such requests each year.**

STAYING HEALTHY

Even though the goal of this book is to help you survive terrorism and other violence, we'd be remiss if we didn't share what we know about protecting your health overseas. Of all the dangers that international travelers face, health threats are by far the biggest. You are, in fact, more likely to die from a disease contracted in a foreign country than from a terrorist or criminal attack. Here, then, are the most significant health issues you need to consider before going abroad.

Be aware of any infectious diseases at your travel destination.

The Centers for Disease Control and Prevention (CDC) provides up-to-date information on international health problems that could pose serious hazards to travelers. Countries infected with quarantinable diseases are listed in the CDC's "Blue Sheet," which is updated every two weeks. The CDC's travel Web site www.cdc.gov/travel/index.htm also has tips on how to prevent insect bites and how to protect yourself against food- and waterborne illnesses (e.g., avoid ice because it could be made from contaminated water). You might also visit Travel Health Online at www.tripprep.com for additional information.

Inspection scores for cruise ships are found in the CDC's "Green Sheet," which is also available at the above Web address.

Get immunized before going abroad.

You can get vaccinated against many travel-related diseases at most state and local health departments. To find the location of the office nearest you, see "Public Health Resources: State Health Departments" at www.cdc.gov/ mmwr/international/relres.html. For information on private travel clinics in the United States and overseas that offer immunizations, contact the International Society of Travel Medicine at P.O. Box 871089, Stone Mountain, Ga 30087-0028, 770-736-7060, www.istm.org, or the American Society of Tropical Medicine and Hygiene at 60 Revere Drive, Suite 500, Northbrook, Il 60062, 847-480-9592, www.astmh.org.

✔ The CDC provides information on traveling globally and returning to the United States with pets and animals at www.cdc.gov/ncidod/dq/animal.htm.

Know who the good doctors are and which hospitals and clinics are safe.

It's wrong to assume that you can get the same quality of medical treatment overseas as you can in the United States. Getting proper medical attention can be especially difficult in developing nations, but even in many of the world's industrialized nations, health care can be hit-or-miss. The problem of inferior medical care around the world therefore requires you to become proactive. You need to look out for yourself by ensuring that you're seen by a qualified physician and, if necessary, admitted to a safe, well-run hospital or clinic.

The U.S. State Department provides the names of qualified doctors and hospitals around the world at http://travel.state.gov/acs.html. The local U.S.

embassy in a foreign country will do the same. (See our contact list in the back of the book for embassy telephone numbers.) Another terrific resource that we recommend is the International Association for Medical Assistance to Travellers (www.iamat.org). IAMAT, a nonprofit group founded in 1960, informs travelers of worldwide health risks, protective immunization, and where to turn when a medical problem arises. Membership is free. Its online directory of physicians directs members to participating physicians, specialists, clinics, and hospitals in 125 countries. IAMAT continuously inspects clinics, hospitals, and physicians' offices around the world and reviews professional qualifications. Its telephone number in Canada is 519-836-0102, and its U.S. number is 716-754-4883.

Recognize that blood supplies overseas aren't always safe.

Although governments are loath to admit it, blood supplies are tainted with disease in more foreign countries than you probably imagine. Until recently, for instance, HIV was a serious problem affecting blood supplies in several Western European countries. In regions such as Asia and Africa, HIV remains a major problem. Blood can also be tainted with hepatitis and other life-threatening disease. Take nothing for granted if you're admitted to a hospital overseas.

Bring copies of drug prescriptions and an extra set of glasses or contact lenses.

Running out of a prescribed medicine while abroad can be a nightmare. Pharmacists won't refill the medication simply based upon the information on the label made out by your local pharmacy. You'll be forced to see a doctor, and that can prove difficult and time-consuming. Make it a point, therefore, to bring copies of all your prescriptions with you. Similarly, if you wear glasses or contact lenses, bring along spares. It's also a good idea to pack any over-the-counter medications that you use regularly (e.g., Tylenol or antacid). Women should further take note when packing that many brands of feminine-hygiene products frequently aren't available overseas and that the products that are on the shelves are often of inferior quality compared with those found in the United States.

✓ If worse comes to worst, have you personal physician back in the States send you a new prescription via overseas courier.

🚫 When returning from overseas, U.S. residents can import up to 50 dosage units of a controlled medication without a valid prescription. But the medications must be declared at Customs upon arrival, must be for your own personal use, and have to be in their original container. Travelers should be aware that drug products not approved by the U.S. Food and Drug Administration (FDA) cannot be brought into the country. Also, the FDA warns that such drugs are often of unknown quality and generally discourages buying drugs sold in foreign countries. For more information, go to the Web sites for the U.S. Customs Service at www.customs.gov and the FDA at www.fda.gov.

Carry adequate medical insurance, and sign up with an air-ambulance service.

Have adequate heath insurance, and be sure the plan covers you when you're overseas. Carry both your insurance policy identity card as proof of insurance and a claim form. Remember, too, that Social Security Medicare doesn't cover medical or hospital expenses incurred outside the United States. Senior citizens, prior to departure, might contact the AARP (www.aarp.org) for information about foreign medical-care coverage with Medicare supplement plans. Another good idea is to sign up with an air-ambulance service, which can whisk you out of a foreign backwater, or indeed any country, and get you to a proper hospital for emergency treatment. A list of air-ambulance services appears at the back of this book. You can find a similar list online at the U.S. State Department, http://travel.state.gov/medical.html. Seniors, in particular, should be aware that some air-ambulance services will provide you with escorts who will fly home with you via a commercial airliner if you're not so desperately ill as to require an air ambulance but do need the assistance of a private helper.

✔ Your life could be taken at any moment in an accident or act of violence, so carry sufficient life insurance to meet the needs of your family. Also, leave a power of attorney with a family member or friend before you head overseas.

🚫 Air ambulances are expensive. A flight could easily run you $10,000. So get special insurance or join a program to cover the expense. Most regular health insurance plans don't cover the cost of air ambulances.

✦ CHAPTER 6 ✦

Airline Safety

FLYING SMART

Airports and jetliners are two primary terrorist targets, yet air travel today is an integral part of our lives—especially our business lives. This has presented most of us with a dilemma: How to balance the need to travel by air against the risk of a terror attack? Well, there are smart and not-so-smart ways to fly. By following some basic procedures, you can cut your risk of involvement in a terrorist incident and increase your chances of survival if you do become an unfortunate victim.

Check ahead for the latest news of any terrorist threat or other security problem.

For special advisories concerning terrorist and other security threats at your travel destination, call the U.S. Department of Transportation's Travel Advisory Line at 800-221-0673. This recorded announcement provides the latest U.S. and overseas travel warnings. (Travelers headed overseas should also call the U.S. State Department's hot line at 202-647-5225.) The Federal Aviation Administration's Air Traffic Control System Command Center at www.fly.faa.gov provides up-to-the-minute information on arrival and departure delays at airports throughout the country. For flight-specific information, you must contact your airline.

Dress sensibly when you fly, giving special thought to your shoes.

Wearing a formal business suit or a stylish dress could prove a hindrance if a quick evacuation of an airplane or airport is required. Causal dress is much more appropriate for air travel these days. Roomy clothing that doesn't restrict motion is advisable. Also, wear clothes made of natural fabrics, such as cotton, wool, denim, or leather. In a fire, synthetics tend to melt. If you aren't carrying luggage in which you can store your business attire, at the very least wear sensible shoes. Carry your dress shoes or high heels in your briefcase or handbag and change into them later.

✔ Dress to cover as much of your skin as possible. This will help to protect you from cuts, abrasions, and skin burn if you have to slide down a plane's emergency chute.

🚫 Avoid sandals, because they could pose a problem if speed and sure-footedness are called for. Women should be aware that nylon stockings can melt, causing injury when sliding down a plane's emergency chute.

Choose paper tickets over paperless ones.

If you have an option, ask for a traditional paper ticket. It will get you through security check-in quicker than an electronic, paperless one. A paper ticket also comes in handy if you need to change airlines because of a flight delay or cancellation. Airlines are usually reluctant to accept paperless tickets issued by another carrier.

Avoid aisle seats, as the chances of injury or death in a hijacking are greater for those passengers seated along the aisle.

Don't brush off the importance of keeping an eye on your bags and knowing whether someone else handled them.

All of us know the drill: we're asked at check-in whether we packed our own bags and whether we left them unattended at any time. Most of us are so bored hearing the questions that we reply with automatic, unthinking answers. Fact is, we're being all too indifferent to the seriousness of the problem and dangerously blasé in answering security questions. Terrorists and criminals have successfully used unwitting passengers to carry bombs or other dangerous items on board aircraft—either by tricking them into carrying packages aboard or by slipping objects into unwatched bags. So think twice before you answer those routine luggage questions the next time you're at a check-in counter. If you have any doubts, be honest and say so. A little white lie in this case could cost you your life.

✓ Don't bring wrapped gifts with you. You may be forced to unwrap them at the security checkpoint, causing you unnecessary delay. Also, before getting to security checkpoints, place in your carry-on luggage any cell phones, pagers, keys, lighters, or other items that will trigger the metal detectors.

Spend as little time at the airport as possible.

Airports appeal to terrorists as sites to attack for a variety of reasons: (1) Airports have high concentrations of people, meaning that any terrorist attack with explosives, gunfire, or biochemical weapons is sure to result in many casualties. (2) Airports are mainstays of modern economies in which business often requires long-distance travel; airport attacks, therefore, have ripple effects throughout an economy in that they make people hesitant to fly. (3) Airports are high-profile targets in that an airport attack anywhere is sure to garner worldwide publicity—publicity that terrorists crave. It therefore behooves travelers to spend as little time in airports as possible. And one of the easiest ways is to avoid checking in luggage.

🚫 Stay away from heavily glassed areas in airports in case a bomb goes off; don't stay in areas outside the security zone; and if you see a disturbance, move away because it could be an unfolding terrorist incident or a ruse staged for thieves to snatch your luggage or wallet.

Whenever possible, don't check any luggage; carry all of your belongings on board.

Being able to carry all your belongings aboard a flight means you will spend less time at the check-in counter and you won't have to go to baggage claim upon arrival at your destination. You should have luggage that will fit either under a plane seat or in the overhead bins. The maximum-size carry-on bag for most airlines is 45 linear inches (i.e., the total of the height, width, and depth of the bag). The Federal Aviation Administration has a hot line, 800-322-7873, open weekdays during normal working hours, designed to answer consumer questions about carry-on baggage, as well as such other issues such as child restraints and turbulence problems. Also, check with your airline to see how many items they allow on board. Carriers normally permit only one carry-on item of luggage and one personal item, like a briefcase or handbag. Next time, pack light. The quicker you're in and out of an airport, the safer you'll be.

✓ If you do check baggage, include sufficient clothing in
your carry-on items to get you through the next day.
You never know when your checked bags will be lost or delayed.

 Don't exchange items between bags while waiting in
line at check-in, security screening, or customs, as this
could raise suspicions.

Never make yourself a slave to your luggage.

Pack light so you can get in and out of travel hubs quickly, without need of
assistance. Having to find a porter is not only a costly and time-consuming
nuisance, it also impairs your ability to avoid trouble at air or rail terminals.
The last thing you want to do these days is to linger at known venues of terrorist attack or criminal activity. Make sure that any suitcase too large or
heavy to carry comes with wheels. And also make sure that your luggage is of
light construction. By that, we don't mean luggage that's going to fall apart,
but suitcases that are constructed of light yet durable materials.

Traveling light usually requires packing well in advance—especially if traveling for several days or weeks. Only by packing well ahead of time can you
determine whether you'll be able to manage without help. Traveling light also
means packing only necessities. That's often easier said than done, however.
The difference between a necessity and a nonnecessity isn't always clear. But
there is a surefire way to tell. Upon returning from your next trip, make it a
point to itemize the contents of your luggage. Look at the clothes you wore
and the items you used—then look at all the rest. You'll likely find plenty of
clothing and paraphernalia that served no purpose and only weighed you
down. This postreturn sorting will help you separate necessities from
nonessentials. Make a list of your travel necessities—and stick to it the next
time you pack for a trip.

✔ Another way to travel light, especially on long trips, is to buy clothes at your destination. Prior to your return, you can mail or ship home items that are too much to carry. Besides, clothes bought abroad or in a fashion city might make nice additions to your wardrobe. Be aware, however, that you may be charged duty on items sent back to the United States from overseas, including clothing bought in the United States, or that you have long owned. For more information, see "Mailing Goods to the United States" at www.customs.gov/travel/travel.htm.

Always try to keep one hand free when carrying your luggage.

Anyone overloaded with baggage is a target for criminals and terrorists. If it takes all you can do merely to keep hold of your belongings, you won't be able to react quickly in a threatening situation. Make it a rule always to keep one hand free when traveling with luggage. That means either packing light or using a luggage rack on wheels.

✔ Have your hotel launder and dry-clean your clothes. That way, you won't have to carry as much. Be sure to check ahead to ensure that your hotel has laundry and dry-cleaning services and find out how long it normally takes to get clothes back.

Carry only product samples that you can handle without assistance.

When the nature of your business requires you to show samples, you want to make traveling with them as easy as possible. The last thing you want is to dally at an airport or a rail station waiting for a porter. So ensure that you can carry both your samples and your personal belongings without assistance. If your sample item is large or heavy, don't bring it with you when you travel. Have it shipped to your destination in advance. You have several options here: (1) ship the samples to your company's office in that city; (2) send them to your hotel; (3) ship them to your customer's office; or (4) have the shipper hold them until your arrival.

Travel in pairs. You and a coworker could handle samples, as well as your luggage, without additional help.

Make any checked luggage easy to spot.

There's nothing worse when waiting for luggage at an airport baggage carousel than to find that someone else is traveling with bags that are identical to your own. You face the prospect that another traveler will mistakenly walk away with your luggage (or vice versa). Moreover, if your luggage is hard to distinguish from everyone else's, you may end up standing at the carousel longer than necessary as your bags make repeated passes before your unseeing eyes.

Adding a distinguishing mark to your luggage solves the problem. Adhere a few decals or masking tape to your bags, or tie something distinctive around the handles. That way you can spot your bags immediately and be out of the airport while your fellow passengers are still straining to find their luggage. Make sure that you can see the marking from a distance.

At an airport baggage carousel, position yourself near the luggage chute.

We don't know if Einstein was the first to postulate this or not, but time takes longer when you're waiting for your luggage. Besides the boredom of waiting, however, the baggage-claim area of an airport isn't a place you want to be these days. Terrorists are always on the lookout for spots where people congregate. At airports, two high-density locations are the check-in counters and baggage claim. The fastest way to get out of harm's way is to retrieve your bags as soon as possible (assuming you need to check any luggage at all). And the easiest method of getting your bags quickly is to stand near the chute from which luggage emerges and slides onto the carousel. This will ensure that you won't have to wait for your luggage to make a slow journey around the carousel to your outstretched hands.

Don't advertise your corporate affiliation on your luggage or ID tags.

Be sure nothing on your luggage identifies your company or your job title. Corporate logos or titles like "Chairman and CEO" will tip off would-be kidnappers and extortionists. Don't use business cards as luggage tags, and always use ID tags that cover your name and address. Provide your name, along with the street address and phone number of your company. Avoid using your home address and number just in case you're abducted and ransom is demanded. In most instances, it's better that your abductors contact your employer as opposed to your family.

> You can increase your chances of retrieving lost luggage if you include identification *inside* as well as outside your bags. Before getting rid of lost luggage, airlines typically look inside to see if they can identify the owner.

Beware of anything you find on a plane, however innocuous it might appear, that could contain a bomb.

If you find what appears to be forgotten luggage in an overhead bin or a piece of electronic equipment, such as a CD or cassette player, left on the floor or wedged in or under your seat, report it immediately to a flight attendant. It could well be a bomb planted by terrorists.

If seated next to an emergency door, take your responsibility seriously.

Emergency evacuations are serious, and seats next to emergency doors carry a heavy responsibility. Moments can mean the difference between life and death in a crashed and burning plane. In the event of a hijacking, furthermore, aircraft emergency exists could be your route to freedom. So take the time to read the instructions on how to open an emergency door and unfurl the slide. Lives could depend on it. If you aren't physically or emotionally up to the task, ask for another seat.

Don't race down aircraft aisles or make any move toward the cockpit.

With air marshals now aboard flights, especially ones bound for such high-security destinations as Washington, D.C., and New York City, any unusual moves around the cabin could get you attention you neither need nor want. Who wants to be tackled and handcuffed when you only intended to beat a hasty path to the lavatory? Think about your movements when aboard an aircraft (and at airports). Don't do anything that might raise suspicions, like deciding to stroll up to the cockpit door.

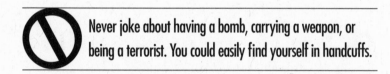

Never joke about having a bomb, carrying a weapon, or being a terrorist. You could easily find yourself in handcuffs.

AIR TRAVEL SECURITY RULES

Following the tragedy of September 11, Congress enacted and President Bush signed new aviation security legislation, which, among other things, established the Transportation Security Administration (TSA). The agency's mission is to protect the U.S. transportation network—most especially, airlines and airports. The TSA, for example, is the organization responsible for the new list of items that air travelers are barred from bringing into aircraft cabins. These new rules, as well as other TSA advisories, are worth reviewing.

Avoid airport hassles by leaving barred carry-on items at home or putting them in your check-in luggage.

Go to the TSA's Web site, www.tsa.gov, for the latest dos and don'ts of air travel, including items allowed in cabins and in checked luggage or not allowed aboard aircraft at all.

Don't put undeveloped film in your checked luggage, because the screening equipment will damage the film.

Review the TSA's guidelines for boarding flights.

Besides allowing yourself extra time for airport check-in—approximately one to two hours—review beforehand the following TSA guidelines for boarding flights in the United States.

Check-in:
- A government-issued ID (federal, state, or local) will be requested. Each traveler should be prepared to show ID at the ticket counter and subsequent points, such as at the boarding gate, along with an airline-issued boarding pass.
- Curbside check-in is available on an airline-by-airline basis. Travelers should contact their airline to see if it is available at their airport.
- E-ticket travelers should check with their airline to make sure they have proper documentation. Written confirmation, such as a letter from the airline acknowledging the reservation, may be required to pass through a security checkpoint.

Screener checkpoints:
- Only ticketed passengers are allowed beyond the security checkpoints. (Arrangements can be made with the airlines for nontravelers accompanying children, and travelers needing special assistance to get to the gate.)
- Don't discuss terrorism, weapons, explosives, or other threats while going through the security checkpoint. Don't joke about having a bomb or firearm. The mere mention of words such as *gun, bomb,* etc., can compel security personnel to detain and question you. They are trained to consider these comments as real threats.
- Each traveler will be limited to one carry-on bag and one personal item (such as a purse or briefcase). Travelers and their bags may be subject to additional screening at the gate.

- All electronic items (such as laptops and cell phones) are subject to additional screening. Be prepared to remove your laptop from its travel case so that each can be x-rayed separately.
- Limit metal objects worn on your person or clothing.
- Remove metal objects (such as keys, cell phones, change, etc.) prior to passing through the metal detectors to facilitate screening. (Putting metal objects in your carry-on bag will expedite going through the metal detector.)

At all times:
- Control all bags and personal items.
- Do not accept any items to carry on board a flight from anyone unknown to you.
- Report any unattended items in the airport or on an aircraft to the nearest airport, airline, or security personnel.

The TSA further says that if you have photo identifications for your children, have those with you and also bring a printout of your e-ticket itinerary. For further information, contact the Transportation Security Administration, 400 Seventh Street SW, Washington, DC 20590, 866-289-9673, www .tsa.gov. Consumers with concerns or complaints about airline or airport safety or security should call the Federal Aviation Administration at 800-255-1111 or go to www.faa.gov.

> ✔ The TSA advises passengers to avoid wearing shoes, clothing, or other items that contain metal, because these will set off airport metal detectors.

YOUNG OR DISABLED AIR TRAVELERS

Recognizing that the new airport security procedures can make travel with young children or travel by persons with disabilities or special needs more difficult, the aviation authorities, among others, have issued some special guidance.

Build children's confidence by putting security procedures in a positive light.

Pessimists view a glass as half-empty; optimists see it as half-full. Attitude, in other words, can have a large bearing on how young travelers respond to heightened security at airports and elsewhere. Parents should, therefore, talk with their children about the new security procedures before getting to the airport, putting an accent on the positive and eschewing the negative. Point out how security keeps us safe and helps us to get where we're going. Don't speak of dangers or threats, and don't discuss September 11 unless *they* bring it up and then explain that that won't happen to them or to you. You might reassure them that the bad men who did that are dead.

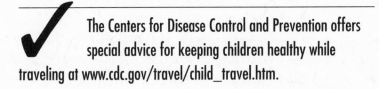

The Centers for Disease Control and Prevention offers special advice for keeping children healthy while traveling at www.cdc.gov/travel/child_travel.htm.

Heed the TSA's advice when traveling with children.

"Before entering the line for the passenger checkpoints," the Transportation Security Administration recommends, "parents should discuss the entire process with their children so they will not be frightened or surprised. Parents should advise children that their bags will be put in the X-ray machine, but it will come out at the other end and that someone may ask to see mom's shoes but they will return them, etc."

Parents should talk to their children before coming to the airport and let them know that it's against the law to make threats such as "I have a bomb in my bag," the TSA warns. Threats made jokingly (even by a child) could result in law enforcement being summoned to the security checkpoint, the entire family being delayed, and eventually being fined or arrested.

For further information, contact the Transportation Security Administration, 400 Seventh Street SW, Washington, DC 20590, 866-289-9673, www.tsa.gov.

Passengers with disabilities or special needs should seek assistance at airports.

In light of stepped-up airport security, special provisions are being made for passengers with disabilities or special needs, particularly during screening. For further information, contact the Transportation Security Administration, 400 Seventh Street SW, Washington, DC 20590, 866-289-9673, www .tsa.gov. If you experience a problem or have a complaint, contact the U.S. Department of Transportation's Aviation Consumer Protection Division, which has a 24-hour complaint line, 202-366-2220 (TTY 202-366-0511), or go to www.dot.gov/airconsumer/.

Find out about accessibility at your travel destinations.

The U.S. Architectural and Transportation Barriers Compliance Board (aka Access Board), a federal agency, offers a variety of free publications for travelers with disabilities. Contact the board at Suite 1000, 1331 F Street NW, Washington, DC 20004-1111, 800-872-2253 or 202-272-0080 (TTY 800-993-2822 or 202-272-0082), www.access-board.gov.

FOREIGN AIRLINE SAFETY

The safety records of many foreign airlines, especially in the third world, are abysmal. It frankly isn't safe to fly on most of the world's airlines. Of the 210 or so countries and other domains that occupy the globe, only 77 are known to meet minimum air safety requirements. That's roughly one in three. It's not a ratio that inspires confidence. Therefore, unless you use a knowledge-able travel agent to book your flights, you need to do some research to ensure your safety when using carriers other than U.S.-based airlines while traveling overseas.

Verify that any foreign airline you use meets U.S. safety standards.

Air travel is the safest in those countries that have had their air systems reviewed and approved by the U.S. Federal Aviation Administration. The FAA, in effect, rates airline service around the world as either safe or unsafe. The FAA doesn't say this in so many words, however. It doesn't, for instance, tell you whether a particular foreign airline is okay to fly. But you can deduce that from information that the FAA supplies under something called the International Aviation Safety Assessment Program (IASA). Under the program, begun in 1992, the FAA assesses the civil aviation authority of each country with service to the United States. These civil aviation authorities are the FAA's equivalents abroad. The FAA's assessments determine whether a civil aviation authority overseeing airline operations to and from the United States meets the safety standards set by a United Nations body known as the International Civil Aviation Organization (ICAO).

The FAA has two ratings for the status of a foreign civil aviation authority: (1) it complies with ICAO standards, or (2) it doesn't comply with the standards. To be in compliance means a civil aviation authority has been assessed by FAA inspectors and has been found to license and oversee air carriers in accordance with ICAO aviation safety standards. When a country isn't in compliance, it means the FAA has determined the civil aviation authority doesn't meet minimum safety oversight standards.

The latest FAA safety ratings for some 100 countries can be found online at www.faa.gov/apa/iasa.htm or by calling the FAA's weekday, working-hours number at 800-322-7873. The FAA also has a 24-hour safety hot line at 800-255-1111, which is primarily used to report safety violations, although you can speak with an FAA representative to discuss a time-critical safety issue.

 Try to fly wide-body planes, because terrorists often avoid hijacking them.

Factor in the risk of terrorist retaliation against the airlines of countries at war with terrorism.

In the aftermath of September 11, the United States formed a rather loosely knit global coalition to fight terrorism. Some countries have taken this commitment more seriously than others. Britain, for example, launched missile strikes against Afghanistan to coincide with the U.S. assault in the fall of 2001; later, British and Canadian troops fought alongside U.S. troops in the region. Other countries, such as Belgium, France, Germany, Israel, Pakistan, Singapore, and Spain, have rounded up terrorist suspects, and still others have cut off the terrorists' money supply.

Terrorists may decide to retaliate against members of the global antiterrorism coalition—perhaps by hijacking or bombing their airliners—but it's impossible to say which countries would be at the most risk. Nonetheless, when considering flying on an airline based in a foreign country, the prominence that country has had in the war against terrorism is worth noting.

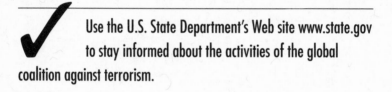 Use the U.S. State Department's Web site www.state.gov to stay informed about the activities of the global coalition against terrorism.

HIJACKING SURVIVAL

The four plane hijackings of September 11 were unusual in two respects: (1) The hijackers took over the planes using box cutters and other small knives, although bombs were also mentioned. Hijackers more typically use guns. (2) The hijackers never intended to land the planes. Most hijackings are used as means of gaining prisoner releases and on occasion ransom money. With the concerted effort to increase air-travel security since the carnage of September 11, it's impossible to say what tactics hijackers will employ in the future. But don't expect terrorists simply to abandon this favored meaning of inflicting damage and striking fear into their enemies.

Know what to expect in a plane hijacking and what to do during a commando rescue operation.

Familiarize yourself with what happens in a typical airplane hijacking so you won't be taken by complete surprise should you ever become a victim. You can expect to hear a lot of noise, commotion, and movement around the plane, even though you may not be in a position to see any of the activity. There could be shooting. You may hear the hijackers yelling at passengers, telling them to stay seated, not to move, and to be quiet. You might even hear an announcement over the plane's public-address system, given by a hijacker or crew member, informing you of the hijacking, making bogus claims about being diverted, or simply saying nothing is wrong.

Hijackers will often force passengers to take new seats to pack everyone together. In some instances, hijackers have separated passengers by religion, race, sex, or citizenship. Your passport and wallet may be taken from you. Overhead luggage bins may be ransacked and seats scoured for cell phones, computers, briefcases, and purses.

If the plane lands, passengers and crew may be used as bargaining chips. Some may be released in exchange for food and water. Others might be executed. Appreciate that hijackers aren't viewed as criminals in every country in the world. The hijackers and the government of the country in which you land may share the same political ideology. Local officials may have more sympathy for the terrorists than for you and the other passengers. So don't automatically expect rescue efforts to be made. But don't become visibly angry because the hijackers may single you out and make an example of you to your fellow passengers.

If you land in a country that opposes terrorism, don't expect your plane to take off again. Authorities will do everything to keep your aircraft on the ground, including blocking runways, denying aviation fuel, and even shooting out the tires. You can expect a rescue attempt to be made—and it's likely to be violent. The plane's doors could be blown open; deafening stun grenades, emitting thick smoke, might go off; and gunfire could erupt. So stay down, placing your head low to the floor. Follow any instruction rescue commandos give you. You might be told, for instance, to stay down or to crawl to the emergency exits. If instructed to exit the plane, you may want to raise your hands in the air as if surrendering; whether this is necessary depends on the specific situation. Should a fire break out, quickly head for the nearest exit.

It's vital that you say and do nothing if you notice a rescue effort being mounted—for example, if you see commando forces, perhaps dressed all in black, climbing onto the wings, running beneath the plane, or propping ladders up against the fuselage. Don't give a hint of what you see to anyone, even the passengers seated next to you. Surprise is essential to a successful rescue, and you don't want to do anything that might tip off the hijackers.

Don't be surprised or insulted if your rescuers treat you at first as if you're one of the hijackers. It's procedure. Security forces need to quickly determine whether any hijackers are attempting to disguise themselves as passengers. Finally, if a hijacking ends overseas, U.S. officials will help you get home. Expect to be debriefed by law enforcement and perhaps U.S. intelligence agents.

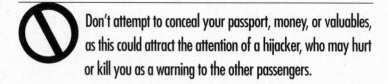

✔ If a hijacker grabs the passenger next to you to use as a shield during a commando raid, try to pull her back. A woman's life was saved that way in March 1991 when commandos of the Singapore armed forces retook a hijacked Singapore Airlines plane.

🚫 Don't attempt to conceal your passport, money, or valuables, as this could attract the attention of a hijacker, who may hurt or kill you as a warning to the other passengers.

Never look your abductor in the eye.

The number one rule in a hijacking, or any type of abduction, is to never look your abductor (or abductors) in the eye. Direct eye contact intimidates a hostage taker and creates in him a fierce animosity toward you. He'll single you out as a troublemaker and may make an example of you to the other passengers.

Don't try to be a hero in the first five crucial minutes of a hijacking.

Your behavior during the first five minutes of a hijacking (or kidnapping) will largely determine how you'll come out of it in the end. In those initial minutes, a terrorist is nervous and perhaps scared. Even though he planned his action, the reality of it means he's facing a lot of unknowns as events unfold. This makes him uptight and trigger-happy. Anything that adds to his fears could lead to disaster. In the first few minutes of a hijacking, therefore, do everything your abductor tells you to do you. This isn't the time to react to the hostage-taking, unless you're trained for it.

Security professionals, such as sky marshals, know that the first few minutes of a terrorist incident are preciously the time for them to counterattack, because the perpetrator isn't set yet, he's still nervous, and he's uncertain. Trained personnel will use this window of opportunity to take over the situation. They don't want to wait until the terrorist becomes set and relaxed, because then he's in control and security personnel may no longer be in a position to respond. The abductor may tie up his victims or corral them. Advantage, abductor. Most victims of hijackings, of course, aren't trained security personnel. They're average people traveling on business or vacation. So don't try to be a hero in the first five minutes of a hostage crisis. Keep uppermost in your mind that yours isn't the only life at stake. You're likely one of a number of hostages. A foolhardy action on your part—particularly during those first crucial minutes of an attack—could endanger the lives of everyone on board. So sit tight, be quiet, and don't make a move. Say a prayer if that's agreeable to you.

A hijacker begins to let his guard down after the first five minutes of the attack. Why? Because he's physically and emotionally spent. The adrenaline rush that came with the initial assault has worn off, and he has become physically and mentally exhausted. He's now less alert and slower to respond. After the first few minutes, be on the lookout for ways to overtake or flee your hijacker. A group of physically fit passengers could well take him down. You might start with a barrage of shoes, thrown at the hijacker by everyone in close proximity, as others passengers rush him. To frighten and distract him, everyone should yell. However, you must be careful and weigh the consequences. People could be shot and killed. If the hijacker actually has a bomb, the entire jetliner could be destroyed. At high altitude, a bullet through a window or the fuselage could lead to rapid depressurization of the plane. And if your hijacker is a well-trained, professional terrorist, your odds of success are low. Still, in some situations passengers can defeat a hijacking. Keep that in mind. But, again, never do anything during the first five minutes of a hijacking.

Appear fully cooperative and observe your attacker.

Look passive and cooperate fully as the hijacking unfolds. Put your head down. Don't ask questions. Don't yell or shout. Don't do anything. Be controlled and reserved. Tell yourself to be calm. If a terrorist addresses you, answer in a normal tone of voice. Observe, out of the corners of your eyes or with quick glances, the hijacker's actions and reactions. Watch to see exactly what he's doing, without making eye contact.

Don't speak unless spoken to.

Don't address your captors unless you're spoken to first, and then keep your answers brief. Never volunteer information to the hijackers, and try not to tell them anything that could help them.

 Avoid being seen talking with fellow passengers, and don't overtly signal or shout to relatives, friends, or colleagues seated elsewhere on the plane.

Don't try to negotiate or reason with your abductors.

You may be a hostage, but that doesn't all of a sudden make you a professional negotiator. That's not your job. Don't try to reason or negotiate with your abductor. For one thing, the incident has just happened. Your hijacker (or kidnapper) has just carried out his planned assault; he's not going to let you go. He can't, no matter what you say. Neither is he in a frame of mind conducive to fruitful discussion, most especially not with the people he has just taken captive. Your abductor needs time to calm down and get his head straight.

🚫 **If you or a fellow passenger requires help, get the attention of a crew member. Don't speak to a hijacker unless spoken to or if he has already rendered similar assistance to other passengers.**

Develop a mental picture of the hijacker.

Try to get a good description of your abductor (or abductors). It will assist authorities later. But it may also help you to escape. Gauge his size, height, and age; note his other physical characteristics (e.g., hair color), and look for any scars or tattoos. Try to determine his nationality, paying particular attention to any use of foreign words. If he addresses a compatriot by name, remember it. To determine whether you have a chance of escape, sense if he's an amateur or a professional terrorist. An amateur hijacker will usually wave his weapon around, yell and scream, and perhaps propagandize. He'll be highly nervous, and he may also act erratically. A professional, by contrast, will be low-key, calm, relaxed, and in control of himself. He'll speak logically, often in a normal tone of voice. He'll appear to have either done this before or rehearsed it many times. You may have a chance of escape if the hijacker is an amateur, but it's highly unlikely that you'll be able to flee from a trained professional.

If the opportunity to escape unharmed presents itself, take it.

Passengers on hijacked aircraft have been able to open emergency doors and flee when their abductors' guards were down. If the aircraft is large and the number of hijackers few, you might be able to head slowly back to the rear doors while the plane is on the ground. If you know how to open the door and unfurl the slide, you could get away before the terrorists notice. But be careful. You're putting your life and the lives of the other passengers at risk. Still, hijackers aren't perfect. They make mistakes. They can't be in all places at all times. So be on the lookout for opportunities to escape.

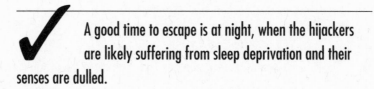

A good time to escape is at night, when the hijackers are likely suffering from sleep deprivation and their senses are dulled.

AIRPLANE EVACUATION

Most travelers have never had to evacuate an aircraft—and probably never will. Still, an understanding of the procedures and knowledge of the precautions you ought to take in the event of a crash landing can significantly improve your odds of survival.

Jump feetfirst onto the emergency chute if you have to evacuate a plane.

Our immediate thought about how to get onto a slide is to sit down, then slide. It likely goes back to our childhood playground days. But in an emergency airplane evacuation, that's the wrong thing to do, because sitting down wastes precious time and could lessen the chances that all of the passengers behind you will have enough time to escape. The Federal Aviation Administration recommends that you place your arms across your chest, elbows in, and put your legs and feet together. Then, jump feetfirst. Don't worry about breaking the chute; they're built to withstand the abuse. Clear the slide immediately, but be careful not to run into any oncoming emergency vehicles. And never return to a burning plane. Also, leave all your possessions behind if you have to evacuate an aircraft. If there's smoke, stay low and follow the floor lighting to the nearest exit.

It's a good idea on any flight to take note of the nearest exits prior to takeoff.

Women should remove their high-heeled shoes and nylon stockings before getting on an emergency slide. A heel could catch on the way down, causing injury such as a twisted ankle or a broken leg. The friction of the slide will melt nylon, possibly causing burns.

Learn a few safety steps that could save your life in the event of a plane crash.

Hardly anyone ever reads the information cards stuffed into seat-back pouches on airplanes, and reading the following tips is no substitute. However, we've come across a number of crash survival tips we'd like to share:

- Count the number of rows between you and the nearest exit. Look both in front of you and behind you. You may not be able to see in a smoke-filled cabin.
- Fasten your seat belt snugly around your pelvic area, not your stomach.
- Pay attention to the flight attendant's safety briefings and instructions.
- Know how to open regular and emergency doors and windows.
- When over water, know where the life jackets and life rafts are stowed, and how to release them.
- Remove all sharp objects, such as pens, from your pockets.
- Get into the "brace" position early. Don't wait until trouble happens. Bend over with your head down, and grab hold of your knees or ankles.
- In an emergency evacuation, leave your personal effects. Just go.
- Crawl on the floor if the smoke is dense.
- Jump over seat tops if you have to.

- Before evacuating into deep water, take your shoes off first. They can waterlog and drag you under.

Consider buying a smoke hood and carrying it with you on all your trips.

Three in four fire fatalities are caused by smoke inhalation. It would be wise, therefore, to consider buying your own smoke hood. These aren't yet standard safety equipment on airlines, but more and more people are making the investment for their own safety. Here are two smoke hoods we've come across: Air Security International's Quick4 is a respiratory-protective escape device designed to reduce the health risks and mortality rate from inhalation of toxic fumes. It costs $159. To order, call 713-430-7300 or go to www.airse curity.com. Brookdale International Systems, a DuPont Canada company, offers EVAC-U8 Emergency Escape Smoke Hood, which combines a DuPont Kapton hood and air filter to protect against toxic smoke while evacuating from fire, chemical, or other emergencies in aircraft, as well as in homes, offices, factories, hospitals, hotels, and marine vessels. It costs $74.95. To order, call 800-459-3822 or 604-324-3822. Or go to either www.evac-u8.com or www.smokehood.com.

3

SAFEGUARDING YOUR IDENTITY AND PROTECTING YOUR BUSINESS

CHAPTER 7

Protecting against Fraud and Identity Theft

Identity theft is much more lucrative than merely stealing wads of cash or using stolen credit cards to buy carloads of goods. If you lost your credit card, for instance, it probably wouldn't take you long to notice and report the problem to your card issuer. However, more than a year usually elapses before a victim realizes his identity has been stolen. And by that time, the thief has likely gotten away with huge sums of money and fraudulently purchased goods and services. Some identity thieves have been known to take out loans to buy cars under other people's names.

Easy access to personal data contained in public records, much of which are available online, is a major concern. In some states and locales, you don't have to be the person of record to gain access to sensitive information, such as a date and place of birth.

"The Internet provides criminals with a tremendous way to locate numerous victims at minimal costs," Thomas T. Kubic, deputy assistant director of the FBI's Criminal Investigative Division, told a congressional panel in May 2001. "The victims of Internet fraud never see or speak to the subjects and often don't know where the subjects are actually located. Crimes committed using computers as a communication or storage device have different personnel and resource implications than similar offenses committed without these tools. Electronic data is perishable—easily deleted, manipulated, and modified with little effort. The very nature of the Internet and the rapid pace of technological change in our society result in otherwise traditional fraud schemes becoming magnified when these tools are utilized as part of the scheme" ("Internet Fraud Crime Problems," Federal Bureau of Investigation, Statement for the Record, Subcommittee on Commerce, Trade, and Consumer Protection, House Committee on Energy and Commerce, May 23, 2001).

Lately, too, a spate of attempts (some successful, some not) have been made to get at hospital records and human resources files. ATM bank-card theft is common, and Dumpster-diving (i.e., going through household trash looking for vital personal data and financial information) is on the rise. And now, with the availability of inexpensive paper shredders, some would-be ID thieves are even piecing together shreds of paper to re-create the original documents.

All of this means, of course, that each of us will have to become increasingly vigilant, looking out especially for signs of financial fraud and identity theft. Here, then, is some helpful information on what to look for and how to handle problems should they arise.

FRAUD AND IDENTITY THEFT

One in three U.S. households have become victims of white-collar crime, including fraud and identity theft, according to the National White Collar Crime Center. Between November 1999 and June 2001, nearly 70,000 Americans reported that their identities had been stolen. The top five states for identity theft are California (17%), New York (9%), Texas (7%), Florida (7%), and Illinois (4%). Traffic in stolen credit cards, meanwhile, is so rife that auction houses, of sorts, have sprung up on the Internet, dealing in long lists of thousands of stolen card numbers.

Be aware of the telltale signs of credit fraud and identity theft.

The trouble with many financial frauds and most identity thefts is the time lag. Victims don't usually discover that they have a problem until long after the criminal activity has occurred. You might, for instance, get a call from law enforcement or be turned down for a loan. Your bank or credit card company may inform you of a problem with your account. Or you may simply discover one day that the money in your checking account is gone.

Here are some of the typical indications that you've become a victim of fraud or identity theft:

- You get a call or letter stating that you have been approved or denied for a credit card for which you never applied.
- You receive credit card, utility, or telephone bills with your name and address, but you never requested any of the services.
- You no longer receive your credit card or bank statements.
- You no longer get mail that you normally expect.
- Your credit card statement includes unauthorized purchases.
- Your bank or credit card company says it has received an application for credit with your name and Social Security number, but you didn't apply.
- You're told by a collection agency that it's collecting for a defaulted account established with your identity, but you never opened opened the account.
- On average, according to the Equifax credit bureau, 14 months go by between the time a person's identity is stolen and the time the victim is made aware of the theft ("Equifax Credit Watch Provides Early Warning of Identity Theft to Consumers," press release, Equifax, Inc., April 10, 2001).

Under the law, the financial institutions involved must bear the brunt of any loss. But the victim is still saddled with damaged credit and must spend months, if not years, regaining his creditworthiness. In the meantime, he may find it difficult to obtain loans, get a job or an apartment, or just write checks.

Before skimping on precautions, appreciate the full cost of identity theft.

Losing your credit, your identity, and even your money isn't the end of the world, of course. Life does go on. But we must warn you of the difficulties you'll face in straightening out your financial affairs, including credit restoration, if your identity is purloined. The task can last years—and cost thousands of dollars—before everything's ironed out.

In 1998, Congress passed the Identity Theft and Assumption Deterrence Act (18 U.S.C. § 1028[a][7]), which prohibits "knowingly transfer[ring] or us[ing], without lawful authority, a means of identification of another person with the intent to commit, or to aid or abet, any unlawful activity that constitutes a violation of Federal law, or that constitutes a felony under any applicable State or local law." The offense, in most circumstances, carries a maximum term of 15 years' imprisonment, a fine, and criminal forfeiture of any personal property used or intended to be used to commit the offense.

Make sense of your wallet and carry only essential money cards and ID with you.

Odds are, if you opened your wallet right now, you'd be surprised at all of the goodies a purse snatcher or pickpocket would inherit. More than likely, your wallet's contents include at least one major credit card, a bank card, and your driver's license. But it's also likely that you've been carrying around charge cards for stores you haven't been to in ages, other credit cards you don't use anymore because they've expired, a PIN (personal identification number) written down somewhere, and perhaps your Social Security card. In fact, you may have put your Social Security card in your wallet so long ago that you can't remember why you needed it in the first place.

A stash like that is a thief's dream. He can now go about stealing your identity without much trouble, because he has your name, address, date of birth, and Social Security number.

Isn't it time you straightened out your wallet? Remove all nonessential credit cards and pieces of identification, most especially your Social Security card, as well as anything with your PIN number on it. Make it a rule to carry only essential ID and the bank and credit cards that you use regularly. It should go without saying that your birth certificate doesn't belong in your wallet, purse, or briefcase. If you need to bring it somewhere, be sure to return it promptly to the safety of your home and store it (along with other vital documents) in a secure, fireproof container or safe.

✔ If you are required, for instance, by a check-cashing firm to show your Social Security card, ask whether they'll accept an alternative piece of identification. If not, take your business elsewhere.

Protect your credit and ATM cards and card numbers from misuse or theft.

Always know where your credit and bank cards are. Use common sense and never leave them lying around. Don't leave your wallet or purse unattended at work or in restaurants, health clubs, or shopping carts. And don't leave a wallet, purse, or handbag in open view in your car, even when the doors are locked. And don't forget to get your card back after you've made a transaction. If you hand your card to a cashier, watch carefully to see that it's only swiped through one scanning machine. Dishonest storeowners and crooked employees have been known to install a computerized second reader that can record a card's coded information, which can later be used to engage in fraud and identity theft. If you see your card swiped in more than one device, report it to law enforcement, the district attorney's office, and the Better Business Bureau.

✔ An easy rule to follow when making credit card transactions is to keep your wallet in your hand until your card's returned; that way, it's unlikely you'll leave a store or restaurant without your card.

Be careful when using ATMs, especially at night.

Before using an ATM, especially at night, take a look around for any suspicious persons or activity. If you have any worries, go to another ATM location—preferably one in a well-lighted area where you can be seen. Have your card ready; don't fumble through your wallet or purse in front of the machine. Let no one see you enter your PIN (personal ID number), using your body to block the view of the keypad. Count your money later, and always take any receipts of your transaction with you. Finally, double-check that you have your ATM or credit card before leaving the machine.

Check your ATM receipts to see if your bank account number is printed out; if it is, ask your bank to remove it.

Commit your PIN to memory.

Safeguard any PINs (personal identification numbers) by never writing them down on anything you keep in your wallet or purse. And don't, as some people do, write your PIN on the back of your bank card. Memorize the number.

Review your list of passwords and select better ones if necessary.

People tend to fall into patterns—patterns that criminals and terrorists can easily exploit—when selecting computer or Internet passwords. So don't use a PIN that a thief could easily crack, such as a birthday, the name of a loved one, or the name of your hometown. Intersperse letters and numbers randomly when creating your PIN, and avoid using the same PIN for all your cards and online accounts. Consider these "password tips" from the Chicago Federal Reserve Bank:

Do
- Use alphanumerics (letters and numbers).
- Change your passwords regularly.
- Close online profiles and accounts when no longer used.
- Log off unattended computers and invoke password screen savers.
- Use different passwords for each Internet access and e-mail account.
- Use first letters of a phrase to create passwords you can remember (e.g., I want a new blue SUV in August = iwanbs08).

Don't
- Share your passwords.
- Use actual words from a dictionary.
- Leave your passwords out where they can be seen.

- Use the same password for more than one Web profile/account.
- Use previously used passwords.
- Use easy-to-guess passwords, such as birthdays, anniversaries, nicknames, Social Security numbers, or Web site names

For more information, see the Chicago Fed's "Money $mart" guide to "safekeeping Web profiles and passwords" at www.chicagofed.org.

Beware of postal theft of your credit card information.

Fraud can occur even before your credit card arrives. Postal theft is one of the leading ways that thieves make off with valuable credit cards. Therefore, if the arrival of a new card or your monthly statement is more than two weeks late, the American Bankers Association recommends the following: (1) Contact the U.S. Postal Service to see if anyone has forwarded your mail to a different address. (2) Check with your bank or credit card company to ask if your statement or card has been mailed. And (3) contact the stores and businesses that normally send you monthly bills to check for any changes in your accounts. If you suspect postal theft, contact the postal inspector at 800-372-8347 or 800-ASK-USPS, www.usps.com/postalinspectors, or through your local post office.

Sign any new card you receive immediately, and keep track of the date on which you expect a new or renewed card to arrive. If it's late, call the card issuer.

If requested to return a card in the mail, say, after closing an account, cut it into pieces before putting it in the envelope.

Never give your account number out over the phone.

Unless you've initiated a call, don't give your credit card or bank account number out over the phone—and even then only provide it if you're certain of the integrity of the person or company you're dealing with. It's a virtual certainty that if you receive an unsolicited call and the caller eventually requests your credit or bank card number, it's a scam. Con artists have even been known to call and say they represent your bank and need your PIN number.

✔ When using a credit card online, make sure the Web site is secure. You can tell if a site's secure by the small key or lock symbol that appears at the bottom of your browser window.

Know how to handle credit card receipts to avoid fraud.

When signing a credit card receipt, draw a line through the blank space above the total so the amount can't be changed later. Never sign a blank receipt. Tear up any carbons before handing the slip back. And save all your receipts to check against your monthly statements. Moreover, never put your address, telephone number, or driver's license number on a credit card slip; such a request should be treated with suspicion.

Inspect your bank accounts for tampering.

Reconcile any ATM or online financial transactions with your monthly bank statement, and call your bank immediately if you discover a discrepancy. It's also wise to keep your checkbook balance up-to-date and call your bank's automated service line regularly to check on your latest transactions and confirm your balance. Write down the date of these calls in your checkbook. Report any irregularities to your bank. Also, store your canceled checks and bank statements in a safe place.

Similarly check your telephone bills for any unusual charges or irregularities.

Be protective of your Social Security number.

Among the vital statistics needed to pull off identity theft, your Social Security number (SSN) is probably one of the more difficult to acquire. That's for a couple of reasons: (1) most people don't carry their Social Security cards around with them, and (2) most people are naturally wary about giving their Social Security number out. Think before you give out your SSN and ask yourself, Does this person really need my number? Why do they want it? Have I ever given it out before to a company like this one or in this kind of transaction? Your best bet is simply never to give your SSN to anyone over the telephone, unless you know for certain that they're on the up-and-up. Usually, companies can wait to get that information, so tell them that you'll send it to them in the mail. That will buy you time to verify the authenticity of the request and the company's credentials.

🚫 Never select the option, offered at some state motor vehicle departments, of using your Social Security number as your driver's license number. If you've already chosen that option, change it and get a new driver's license.

Inform law enforcement and government attorneys of financial scams.

If someone tries to scam you or fails in a bid to steal your identity, report the incident to local law enforcement, your district attorney, state attorney general, and local Better Business Bureau. You can never tell. Your report might be the one that finally nails this person or group. In this new age of terrorism, volunteering information about attempted and realized financial fraud is something of a patriotic duty. Who knows what heinous plot you might foil with one simple phone call. Telephone numbers of state and district attorneys general can be obtained from directory assistance or the National Association of Attorneys General at 202-326-6000 or www.naag.org. And don't feel embarrassed. The National Fraud Information Center (NFIC) reports that telemarketers alone defrauded nearly 39,000 people out of a total of $5.7 million in 2000, with an average loss of $1,462. The most popular phone scams involved prizes, magazine and credit card sales, work-at-home offers, and "advance-fee loans." Internet fraud, including online-auction and general-merchandise scams, took nearly 7,000 people for $4.4 million in phony sales in the first 10 months of 2001. The average loss per person was $636, up from $427 in 2000 ("Telemarketing Fraud Statistics 2000," National Fraud Information Center, National Consumers League). You can report fraud to the NFIC by calling 800-876-7060 or online at www.fraud.org.

Contact the FTC for the best information on card fraud and identity theft.

For more information or assistance regarding card fraud or identity theft, the place to go is the Federal Trade Commission. Its Web site, www.ftc.gov, is chock-full of information, or you can call the FTC hot line, 877-FTC-HELP (877-382-4357).

PROACTIVE MEASURES

Because it often takes considerable time for a victim to become aware of a financial fraud or identity theft, it's best to take prophylactic steps to protect yourself.

Order copies of your credit report.

At least once a year, order copies of your credit report from the three major credit-reporting agencies—Equifax (www.equifax.com), Experian (www.experian.com), and TransUnion (www.transunion.com). Upon receipt of the reports, check for any unauthorized activity. If you discover information in your credit file that doesn't pertain to you (e.g., credit card or bank accounts you never had, delinquent-payment notices, applications for credit, etc.), inform the relevant creditors that you suspect attempted fraud and/or identity theft.

Consider subscribing to an identity-protection service.

Several useful services will monitor your credit files daily for possible fraud and identity theft. They normally inform you within 24 hours of any suspicious activity involving credit associated with your name. The cost of these services ranges from $50 to $90 a year, but at the same time they do save you the trouble and expense of having to contact credit agencies annually. Most important, having your credit files constantly monitored means that you're likely to spot fraud or identity theft quickly, thus limiting any damage an impostor might do to your credit standing. Identity Guard (www.identity guard.com) is one monitoring service, offered by TransUnion and Intersections Inc. Another is Equifax Credit Watch (www.equifax.com), and a third, called Credit Manager, is offered by Credit Expert (www.creditexpert.com) in conjunction with Experian.

Give out your mother's maiden name sparingly.

We were shocked some time ago when a renowned newspaper, which should have known better, asked us to provide our mother's maiden name over an unsecured Internet connection as part of an online subscription. Of all your means of self-identification (e.g., birth certificate, Social Security number, or driver's license), your mother's maiden name is your most valuable and should thus be the one that you guard most dearly. Your mother's maiden name serves as sort of a universal password to verify that you are who you say you are.

Absolutely never give out her maiden name before first thinking through the reason for the request to determine whether it's legitimate. In the main, there are only two reasons you should provide your mother's maiden name to anyone: (1) to open a new financial account, attend a school, or get hired for a job, and (2) to respond to a request from an institution with which you have an account, educational tie, or job that needs to check your identity. Consider any other request for your mother's maiden name, especially one made over the telephone during an unsolicited call, as an attempt to steal your identity.

Maintain only a minimum number of credit cards.

Carrying around a multitude of credit cards is dangerous, so first rid your wallet or purse of unneeded cards. Then rethink the number of credit card accounts you maintain. View each of your credit cards as an opening for a thief, and minimize your risk by reducing the number of your credit accounts. Begin by canceling all unused cards.

Destroy unsolicited, preapproved credit cards.

Preapproved credit card applications are an easy way for someone to steal your identity. They merely apply for the cards and then request a change of address. Worse, once they have the card, they can wield it to apply for others—depending on what *your* income is. This can turn into a torrent of fraudulent credit requests, ranging from auto loans to cellular phone services. You can have your name removed from credit card prescreening programs, which are operated by the nation's three credit reporting bureaus, by calling 888-5OPTOUT (or 888-567-8688). One call will notify all three credit bureaus and should eliminate this source of junk mail. Otherwise, contact them individually: Equifax, 800-525-6285, www.equifax.com; Experian, 888-397-3742, www.experian.com; and TransUnion, 800-680-7289, www.transunion.com.

Reduce the personal information about you that's in circulation.

Beyond eliminating unsolicited credit card offers, various law-enforcement authorities also recommend that you reduce the amount personal information about you that's in circulation by taking the following steps: Sign up with the Direct Marketing Association's Mail Preference Service and Telephone Preference Service (www.the-dma.org or Box 643, Carmel, NY 10512) to get your name deleted from junk-mail, e-mail, and telemarketing lists. And remove your name and address from telephone books and reverse directories; begin by contacting your telephone company.

Personal checks should carry only your name.

Most people provide far too much information on their personal checks. Your checks should carry only your name. Leave your address and telephone number off the next batch of checks you order. That information is immensely valuable to a thief and can too easily be acquired by simply getting hold of one of your checks for a moment or two.

Before doing business with an Internet bank, make sure it has a legitimate bank charter.

Depositing money in a phony bank is one way of being swindled. It's the sort of fraud that can really occur only via the Internet. It's hard to imagine a bank in your neighborhood being there one day and gone the next, but the Internet makes such bold schemes possible. Every legitimate bank has a banking charter and is federally insured through the Federal Deposit Insurance Corp. Check with the FDIC (www.fdic.gov) or the Federal Reserve (www.federalreserve.gov) before opening an online bank account.

Be careful what you throw in the trash, both at home and in public.

If you close a checking account, don't just toss the unused checks away. Carefully destroy them before discarding, paying particular attention to the account number; be sure that the number is at least torn in half and deposit the halves in different garbage bags. Then, don't set these out all at once for trash pickup. Distribute the scraps of paper over different pickup days. Do the same for old bills, canceled checks, credit card receipts, bank statements, mutual fund records, and any financial document that contains vital information about you, such as your birth date and Social Security number.

At restaurants, always take your credit card receipts with you and never throw them into a trash container on the street. Also, once you've signed your credit card receipt, rip out and tear up any attached slips of carbon. After using an ATM, wait for the receipt and take it with you. Never leave it behind, and don't simply throw it in the trash. Bring it home and keep it with your other financial receipts.

If cyberterrorism disrupts the banking system, your receipts could prove more valuable than you think.

Buy a paper shredder for your home.

Take precautions against "Dumpster-diving." That's when thieves go through garbage looking for valuable information. Buy a paper shredder, especially if you do a lot of financial paperwork at home—or have a home office—or discard a lot of job-related materials at home. Even if you tear important documents up into pieces, a determined thief who's after your identity won't be deterred. It may take him time, but he'll eventually be able to put enough material together to impersonate you, open up accounts in your name, and then vanish before you are even aware of what has happened.

 Use paper shredders to destroy unwanted credit card solicitations that come in the mail.

🚫 Only shred the unwanted contents of your mail. Never shred envelopes or package coverings, because they could become contaminated with a biological agent should an outbreak like the anthrax-letter campaign of 2001 reoccur. Discard envelopes by following the mail-handling safety steps laid out in chapter four.

Use regular U.S. Postal Service boxes to mail checks.

An erect red flag on your mailbox is an invitation to a thief, who'd like nothing better than to get his hands on one of your personal checks. So when mailing payments by mail, use the Postal Service's regular mailboxes and not the one at the end of your driveway.

Print out copies of all online financial transactions.

Whether you trade stocks or pay bills online, print out hard copies of all the financial transactions and then store them in a safe place. The reason is simple: Should cyberterrorists attack financial institutions and their databases, records of online transactions could be compromised. Printouts attesting to stock trades, bill payments, and the like would serve as backup verification.

LOST OR STOLEN CARDS

Identity assumption in check fraud occurs, says the U.S. comptroller of the currency, when criminals learn information about a bank customer (e.g., name, address, bank account number, account balance, Social Security number, employer, and home and work phone numbers) and use the information to misrepresent themselves as the real bank customer. The thieves may then alter account information, create fictitious transactions, or draw money out.

Report the loss or theft of credit or bank ATM cards as quickly as possible.

Report the loss or theft of a credit card or bank ATM card to the card issuer as quickly as you can because the timing of your report will affect your financial liability. Many companies have toll-free numbers with 24-hour service to handle emergencies. Then follow up with a letter, giving them your account number, the approximate time and date you first noticed your card was missing, and the date on which you called to report the loss.

Keep a list of all your account and card numbers in a secure location.

It helps to have a list (kept in a safe place) of all your account and card numbers, including card expiration dates and the telephone numbers of the card issuers. That'll speed up the process of reporting any lost or purloined cards. Take this information with you when traveling, but never keep the list in your wallet. It won't do you much good if the list gets stolen along with everything else in your wallet.

Subscribe to a credit card registration service if you hold multiple cards.

For a $10 to $35 annual fee, you can subscribe to a card registration service that will notify the issuers of a theft or loss of your credit cards and bank cards. You need make only one phone call to report all lost or stolen cards, instead of having to contact each card issuer individually. Most service companies will request replacement cards at the same time. Purchasing a card registration service may be convenient, but it isn't required to limit your exposure to liability. Federal law gives you the right to contact your card issuers directly in the event of a loss or suspected unauthorized use. If you decide to sign up for a registration service, compare offers. Carefully read the contract to determine the company's obligations and your liability. For example, be sure the company will reimburse you if it fails to notify card issuers promptly.

Know how to limit your loss from stolen or lost credit and ATM cards.

The Fair Credit Billing Act (FCBA) and the Electronic Fund Transfer Act (EFTA) offer procedures for you to use if your cards are lost or stolen. You also may want to check your homeowner's insurance policy to see if it covers your liability for card thefts. If it doesn't, some insurance companies will allow you to change your policy to include this protection.

The following procedures are recommended by the U.S. Federal Trade Commission (FTC) to deal with financial losses resulting from lost or stolen credit and ATM cards:

CREDIT CARD LOSS   If you report the loss *before* the cards are used, the FCBA says the card issuer cannot hold you responsible for any unauthorized charges. If a thief uses your cards before you report them missing, the most you will owe for unauthorized charges is $50 per card. This is true even if a thief uses your credit card at an ATM machine to get cash. However, it's not enough simply to report your credit card loss. After the loss, review your billing statements carefully. If they show any unauthorized charges, send a letter to the card issuer describing each questionable charge. Again, tell the card issuer the date your card was lost or stolen and when you first reported it to them. Be sure to send the letter to the address provided for billing errors. Don't send it along with a credit card payment or to the address you normally use to make payments, unless you are directed to do so.

ATM CARD LOSS   If you report an ATM card missing *before* it's used without your permission, the EFTA says the card issuer cannot hold you responsible for any unauthorized withdrawals. If unauthorized use occurs before you report it, the amount you can be held liable for depends upon how quickly you report the loss. For example, if you report the loss within two business days after you realize your card is missing, you will not be responsible for more than $50 for unauthorized use. However, if you don't report the loss within two business days after you discover the loss, you could lose up to $500 from unauthorized withdrawals. You risk *unlimited* loss if you fail to report an unauthorized transfer or withdrawal within 60 days after your bank statement is mailed to you. This means you could lose all the money in your bank account and the unused portion of your line of credit established for overdrafts.

If unauthorized transactions show up on your bank statement, report them to the card issuer as quickly as possible. Once you've reported the loss of your ATM card, you cannot be held liable for additional amounts, even if more unauthorized transactions are made. This means, of course, that you must make it a point to open your monthly statements promptly upon receipt and then check the charges against the receipts you've accumulated over the past month. Alternatively, you could make a habit of checking you bank balance more than once a month either online at your bank's Web site or via your bank's 24-hour computerized telephone accounts service. Call the "billing inquiries" phone number on your statement to resolve any discrepancies. Card issuers are required to investigate billing errors reported to them within 60 days of the date your statement was mailed to you, so keep the envelope if it is postmarked.

Report fraud to law enforcement and credit bureaus to begin fraud resolution.

If you become a fraud victim, the first thing to do is call your local police or sheriff's office. But don't use the emergency number. Instead, call the office's regular phone number, which can be found in your phone book or obtained from directory assistance. The authorities frown on frivolous 911 calls. Informing law enforcement of credit fraud isn't something you can avoid, because you'll need a complaint number from police to provide to credit-reporting agencies and your creditors.

Next, report all stolen cards—both credit cards and ATM cards—to the card issuers and request new cards. Follow the call with a written notification, and cite the date and time of your initial call to the company. You'll probably have to fill out affidavits of forgery to establish your innocence with banks and card issuers. Appreciate that you aren't the only victim; these institutions may suffer a financial loss. If you suspect mail theft was involved, contact the postal inspector. Mail theft, by the way, is a felony. Also, contact the Federal Trade Commission to report the incident.

Get in touch with all three of the credit bureaus listed below so you can prevent any further damage to your credit standing and begin credit restoration. Each creditor may have a different process for handling a fraud claim. Make sure you understand exactly what is expected of you, then ask what you can expect from the creditor. Take notes of all phone calls, asking for names, department, and phone extensions; record the times and dates of your calls. At the conclusion of a fraud investigation, ask all involved creditors for a document stating you aren't responsible for the debt.

Credit Bureau Fraud Departments:

Equifax, Consumer Fraud Division, P.O. Box 740256, Atlanta, GA 30374; phone: 800-525-6285 or 404-885-8000; fax: 770-375-2821; www.equifax.com.

Experian, National Consumer Assistance Center, P.O. Box 2002, Allen, TX 75013; phone: 888-397-3742; www.experian.com.

TransUnion, Fraud Victim Assistance Department, P.O. Box 6790, Fullerton, CA 92834; phone: 800-680-7289; fax: 714-447-6034; www.transunion .com.

Government Agencies:

Federal Trade Commission, 877-438-4338, www.ftc.gov.

U.S. Postal Inspection Service, 800-372-8347, www.usps.com/postalin spectors.

Social Security Administration, 800-269-0271, www.ssa.gov.

Identity theft is a federal crime, so report it to Washington.

Identity theft is no laughing matter. It's a federal felony, punishable under the Identity Theft and Assumption Deterrence Act of 1998 (18 U.S.C. § 1028 [a][7]), as well as a crime in roughly half the states, so treat it accordingly. Report any case of identity theft promptly to the Federal Trade Commission, which has a special ID theft hot line, 877-438-4334, or use the form on its Web site, www.ftc.gov.

Seek professional assistance if you've been defrauded.

If you believe you've been defrauded and have questions, contact TransUnion's Fraud Victim Assistance Department (FVAD), which was established in 1992 to detect, prevent, and rectify credit fraud and assist fraud victims. It's staffed by employees trained in detecting and resolving all credit-fraud-related situations; they also can explain the many applicable laws, regulations, and consumer-relations policies. You can e-mail the FVAD at fvad@transunion.com. It will respond to your e-mail as soon as possible. Should you require an immediate answer, however, you may wish to call its toll-free number, 800-680-7289, Mondays through Fridays, 5:30 a.m. to 4:30 p.m. Pacific standard time.

SSN AND CHECK THEFT

Contact SSA if you suspect someone else is using your identity.

Defrauding Social Security has already become big business. In fiscal year ended September 30, 2000, the Social Security Administration's (SSA) Office of Inspector General received nearly 47,000 allegations of misuse of Social Security numbers. These included charges of identity theft, in which stolen numbers were used to collect Social Security benefits or gain employment un-der another person's name. So immediately contact the SSA if you suspect someone else is using your identity by calling 800-269-0271 or 800-772-1213 or going to www.ssa.gov.

Report stolen checks and check fraud to your bank and check verification companies.

If you discover your checks have been stolen or find that checks have been cashed that you didn't write, contact your bank immediately and stop pay-ment. It's further recommended that you close the account immediately and open another one with a new account number. Similarly, if your ATM card has been lost, stolen, or otherwise compromised, cancel the card as soon as you can and get another with a new PIN.

If your checks have been stolen or misused, contact the major check verifi-cation companies to request that they notify retailers not to accept these checks. You can ask your bank to notify the check verification service with which it does business. The major check service companies are National Check Fraud Service, 843-571-2143; SCAN, 800-262-7771; TeleCheck, 800-710-9898 or 800-927-0188; CrossCheck, 707-586-0551; Equifax Check Systems, 800-437-5120; National Processing Company, 800-255-1157; and International Check Services, 800-526-5380. You'll probably have to fill out affidavits of forgery to establish your innocence with your bank and any recipients of stolen checks. These institutions are joint victims with you and could face financial loss. Also, contact the U.S. Postal Inspection Service at 800-ASK-USPS or www.usps.com/postalinspectors/welcome.htm.

✔ To safeguard delivery of your checks, especially if your mailbox doesn't lock or if you live in a high-crime area, tell your bank you don't want your checks sent to you in the mail and that you would prefer to pick them up at your bank branch.

Get a new driver's license number if yours has been misused.

If someone has been using your driver's license number to cash bad checks or to engage in other financial fraud, insist that your state motor vehicle department issue you a new license number. You may be asked to prove that you've been damaged by the theft of your driver's license and/or the misuse of your identity. Be persistent if the bureaucrats give you a hard time.

RESTORING YOUR CREDIT

Patience will be required in getting your credit restored. The paperwork can be mind-numbing. But don't be discourteous. Remember that you can get more with honey than with vinegar.

Know how to get your credit restored if you fall victim to fraud.

The moment your identity is stolen, your credit is damaged. Using your name and credit information, the impostor will ring up purchases, withdraw cash, and maybe get a telephone number in your name. This may not last long, but the amounts involved can be huge. And creditors can't help but think you're the person who owes them all that money. It's then left up to you to reclaim your identity and restore your credit. You need to straighten things out with creditors, remove inaccurate information from your credit report, and take steps to prevent any further fraud.

Here are the procedures that TransUnion, one of the nation's leading credit bureaus, recommends you follow:

REVIEW YOUR CREDIT REPORTS   Get copies of your credit reports from all three credit bureaus—Equifax, Experian, and TransUnion—and review them for any unauthorized account and inquiry information. Should any of the information on your credit file not pertain to you, please contact the credit grantor directly and ask about the account or inquiry.

CONTACT CREDITORS   Explain your situation to the credit grantor and ask for an explanation of the procedures for fraudulent accounts or charges.

You may be required to complete an affidavit of fraud and/or send additional documents. These may include a police report, a copy of your driver's license, and documents from other credit card companies confirming the accounts as fraudulent. Once each creditor acknowledges fraud, ask the creditor to send you and all major credit-reporting agencies a letter of confirmation. Trans-Union suggests you keep a log of all phone conversations when dealing with each credit card company and financial institution. Log dates, names, and notes about what you discussed with each company. Follow up with each company and ask about the progress of the investigation. The inquiries shown on your report can remain there for two years. However, inquiries determined to be fraudulent are removed upon that determination.

CONTACT CREDIT REPORTING AGENCIES Contact the major credit reporting companies and request that a protective statement be added to your credit file. Be sure to ask how long the statement will remain on your report. Ask each company if any recent activity appears on your credit file. If so, ask for each name, address, and telephone number of any unauthorized account or inquiry. In addition, ask to receive a copy of your credit file for you to review. Review the reports for any unauthorized activity, and also contact those creditors to ask about the account or inquiry. Keep in mind that each of the major credit-reporting bureaus has different procedures. It is best that you contact each company and ask about the procedures.

CONTACT FINANCIAL INSTITUTIONS Notify your financial institution. Cancel your checking account and request a new account number. If you are unsure about any outstanding checks, stop payment. If you have experienced fraudulent use of your checks, contact one of the companies that collect, report, and investigate returned checks. (See above for the names and phone names of these companies.)

CONTACT THE SOCIAL SECURITY ADMINISTRATION If your Social Security number was fraudulently used, contact the Social Security Administration (SSA) at 800-772-1213 or www.ssa.gov. Social Security numbers may only be changed when proper documentation is submitted to the SSA. Trans-Union doesn't recommend that you change your Social Security number, as this may cause future complications.

COMPLETE A DISPUTE FORM   You may also complete and return a dispute form and attach any documentation from all credit card companies that were victimized, although the form is not required to submit a dispute. Once your dispute form and/or documents are received in its office, TransUnion contacts each creditor involved and verifies the account information you are disputing. The investigation may take up to 30 days, at which time TransUnion sends you an updated copy of your credit file reflecting the results of its investigation.

> ✔ Order Your Credit Report: Once a year, order reports from all of the major credit reporting companies. Check for any unauthorized activity. Should any information on your credit file not pertain to you, contact the creditors and question the account and/or inquiry.

Keep important paper records of your personal accounts and finances.

Keep paper records so you can compare your bank and financial statements against your personal records in the event of a cyberterrorism meltdown. Here's a list of many of the documents to hang on to: all bank and savings account statements; paycheck stubs and records of other income, including deposit slips; brokerage account statements and confirmation receipts that show when transactions took place; credit card statements, canceled checks, and receipts that show when payments and charges were made; records of all loan balances and payments; and records of all regular charges such as utility bills and insurance premiums.

Store important or irreplaceable documents at two different locations.

For whatever reason, you could face the nettlesome problem of having to replace vital documents, some of which might be irreplaceable. So you're best off making duplicates of important papers and storing either them or the originals in a secure place like a safe-deposit box, a fireproof container at a relative's house, or your lawyer's office.

Important Documents Storage

	Location of Original	Location of Duplicate
Birth certificates	_____	_____
Adoption papers	_____	_____
Marriage certificate	_____	_____
Divorce papers	_____	_____
Citizenship papers	_____	_____
Bank accounts	_____	_____
Investment accounts	_____	_____
IRAs & 401(k)s	_____	_____
Deeds & titles	_____	_____
Mortgages & loans	_____	_____
Car titles	_____	_____
Insurance policies	_____	_____
Insured assets data	_____	_____
Wills	_____	_____
Special instructions	_____	_____
Passports	_____	_____
Tax records	_____	_____
Medical records	_____	_____
Fingerprints, etc.	_____	_____
Diplomas	_____	_____

Be able to verify any direct deposits or payments.

If you have your paychecks or other income directly deposited in your bank account or have bills paid automatically, retain all paycheck receipts and billing information so you can contest any mistakes. If you travel abroad or rely on international financial transactions, make sure you have all records needed to resolve potential financial glitches. And if you use a computer software program, print out statements from time to time to use as backup.

✓ Keep enough cash on hand to cover any shortfalls, or hold multiple bank accounts, providing redundancy in the event cyberterrorist activity is directed at individual financial institutions.

Get a printed history from your lenders.

Ask your lenders for printed histories of the payments on your mortgage, car loans, and other debts. Verify the principal and interest paid. Be sure to request this information at least once a year, and know that some institutions charge a fee for these records.

Check that your funds are adequately insured.

Should cyberterrorism bring a U.S. bank or thrift institution crashing down, don't worry; you're probably insured against any loss. The Federal Deposit Insurance Corp. (FDIC), founded in 1933, guarantees your checking, savings, and money market deposit accounts, as well as certificates of deposit, held in FDIC-insured depository institutions, such as banks and thrifts. The basic FDIC insurance limit is $100,000 per institution. A new FDIC online service at www2.fdic.gov/edie/ allows you to check to see if your accounts at any single institution exceed the insurance limit. Individual Retirement Accounts (IRAs) held in FDIC-insured banks are separately insured up to $100,000. However, investment products that aren't insured include mutual funds, annuities, life insurance policies, stocks, and bonds. For more information, call the FDIC at 877-275-3342 or 202-942-3147.

FRAUDS AND SCAMS

When making donations, follow the FTC's "charity checklist."

The FTC has composed the following "charity checklist" to ensure that your donations actually benefit the people and organizations you intend to help.

- *Ask for written information, including the charity's name, address, and telephone number.* A legitimate charity or fund-raiser will give you materials outlining the charity's mission, how your donation will be used, and proof that your contribution is tax deductible.
- *Ask for identification.* Many states require paid fund-raisers to identify themselves as such and to name the charity for which they're soliciting. If the solicitor refuses, hang up and report it to local law enforcement officials.
- *Call the charity.* Find out if the organization is aware of the solicitation and has authorized the use of its name. If not, you may be dealing with a fraudulent solicitor.
- *Watch out for similar-sounding names.* Some phony charities use names that closely resemble those of respected, legitimate organizations.
- *Know the difference between "tax exempt" and "tax deductible." Tax exempt* means the organization doesn't have to pay taxes. *Tax deductible* means you can deduct your contribution on your federal income tax return. Even though an organization is tax exempt, your contribution may not be tax deductible. If deductibility is important to you, ask for a receipt showing the amount of your contribution and stating that it is tax deductible.
- *Beware of organizations that use meaningless terms to suggest they are tax exempt charities.* For example, the fact that an organization has a "tax ID number" doesn't mean it is a charity: all nonprofit and for-profit organizations must have tax ID numbers. And an invoice that tells you to "keep this receipt for your records" doesn't mean your donation is tax deductible or the organization is tax exempt.
- *Be skeptical if someone thanks you for a pledge you don't remember making.* If you have any doubt whether you've made a pledge or previously contributed, check your records. Be on the alert for invoices claiming you've made a pledge when you know you haven't. Some unscrupulous solicitors use this approach to get your money.
- *Ask how your donation will be distributed.* How much will go to the program you want to support, and how much will cover the charity's administrative costs? If a professional fund-raiser is used, ask how much it will keep.

- *Refuse high-pressure appeals.* Legitimate fund-raisers won't push you to give on the spot.
- *Be wary* of charities offering to send a courier to collect your donation immediately.
- *Consider the costs.* When buying merchandise or tickets for special events, or when receiving free goods in exchange for donating, remember that these items cost money and are generally paid for out of your contribution. Although this can be an effective fund-raising tool, less money may be available for the charity.
- *Be wary* of guaranteed sweepstakes winnings in exchange for a contribution. You never have to donate anything to be eligible to win.
- *Avoid cash gifts that can be lost or stolen.* For security and tax record purposes, it's best to pay by check. Use the official full name of the charity—not initials—on your check. Avoid solicitors who want to send a courier or use an overnight delivery service to pick up your donation.

"Many charities use your donations wisely. Others may spend much of your contribution on administrative expenses or more fund-raising efforts. Some may misrepresent their fund-raising intentions or solicit for phony causes," the FTC says ("Charitable Donation$: Give or Take," advisory, Federal Trade Commission, April 1997). Therefore, before you open your checkbook, check out the charity you're considering with these organizations: Philanthropic Advisory Service, Council of Better Business Bureaus, 4200 Wilson Boulevard, Suite 800, Arlington, VA 22203-1838, 703-276-0100, www.bbb.org; BBB Wise Giving Alliance, 4200 Wilson Boulevard, Suite 800, Arlington, VA 22203, 703-276-0100, www.give.org; and American Institute of Philanthropy, 4905 Del Ray Avenue, Suite 300, Bethesda, MD 20814, 301-913-5200, www.charitywatch.org. In addition, most states require charities to be registered or licensed by the state, so check with your state attorney general or secretary of state.

PLANNING AHEAD

Make out a will and letter of instruction.

Making out a will isn't something most of us care to do, because it brings us face-to-face with our eventual demise. Dealing with death isn't easy, but having an up-to-date will is important. It will facilitate the distribution of your assets to your loved ones and charities after you're gone.

There are a lot of misconceptions about wills. Some people think because they don't have much money or tangible assets, a will isn't necessary. They're wrong. Even if the things you leave behind are only of sentimental value, it's best to indicate in advance the persons you'd like to receive them; otherwise, divvying up your belongings could cause friction among family members. If you and your spouse own assets jointly, at death your share will automatically go to the survivor. But what happens to your children if your spouse remarries and fails to designate your share of the assets to them? And what happens if you and your spouse die at the same time? Who will be the guardian of your children? Similarly, people think they can write their own will. Trouble is, such wills often don't meet at the necessary legal requirements.

So bite the bullet, schedule an appointment with a lawyer, and make out a will. You'll be doing everyone a big favor. And while you're at it, write a letter of instruction covering funeral arrangements and such, and if you've of a mind, make out a living will concerning your medical treatment should you become incapacitated.

Get fingerprinted, get your blood sampled, and obtain your dental X-rays.

The horrific nature of an act of mass destruction can make identification of remains difficult. Be prudent and get yourself fingerprinted at your local police station or sheriff's office. Ask your doctor to take a blood sample and have it stored, so your DNA can be checked. Also obtain copies of dental records from your dentist. Planning ahead will reduce the strain on your family should the need for such information ever arise.

Create a financial plan with the help of an estate planner, lawyer, or accountant.

America has developed an "investor class," as economist Larry Kudlow calls it, over the past decade or more, with the explosion in individual retirement accounts, 404(1)k plans, mutual funds, and online trading. This means, of course, that an awful lot of personal savings are invested in the stock, bond, and money markets, not to mention real estate and collectibles. Should tragedy suddenly befall you, your survivors would be much better off if a financial plan were in place that ensured a steady stream of income and kept exposure to taxation as low as possible. In these uncertain times, therefore, it's wise to sit down with financial and legal professionals and plan your estate to meet the goals you'd like to achieve.

Update, if necessary, the designated beneficiaries of your retirement, savings, bank, and investment accounts.

In all likelihood, it has been years since you last looked at the beneficiary designations you made when you opened your various financial accounts. To ensure that your designated beneficiaries are still the correct ones, contact the institutions in which you hold your money and savings and update the designees, if necessary. You should check your Individual Retirement Accounts, 401(k) plans, Keoghs or their equivalents, mutual-fund brokerage accounts, and bank checking and savings accounts.

Secure adequate life and medical insurance, especially if you travel often.

An accident or act of violence, perpetrated by a terrorist or a common criminal, can take a life at any time and at any age. So you're never too young or too old to carry life insurance to help meet the needs of your loved ones. It also pays to have adequate heath insurance, especially if you're frequently overseas, but make sure the coverage extends beyond the U.S. borders.

Keep all tax-related information up-to-date.

As they say, two things are certain in life: death and taxes. What's odd about taxes, though, is that the government wants them paid even after you're dead. Make sure, therefore, that you keep all information needed for tax-form preparation up-to-date. That's especially important for the self-employed entrepreneur or small-business owner, who is often the only one who knows what's going on in her business. If you die, you certainly wouldn't want to add to the pressures on your spouse or children by forcing them to comb through stacks of papers, unmarked files, and assorted receipts in order to file a tax return.

Keep paper copies of all bills, receipts, bank accounts, mutual-fund statements, stock and bond holdings, and other financial transactions.

In case computerized data get destroyed or become inaccessible, keep copies of all your bills and account statements through the end of the year. These will detail payments toward principal, interest, and other charges, plus any outstanding balance, so you have a record of your account—if there is a computer foul-up. Keep all charge receipts, and check them carefully against the billing statement. Report any discrepancies immediately to the card issuer. Keep canceled checks as proof of payments for recent months. If you bank by computer, download your transaction records before the end of the year and store them on a backup disk. If you haven't saved quarterly reports, get copies of account information from your broker or mutual fund company before year-end.

CHAPTER 8

Protecting Your Business and Buildings

The one question a business owner or executive never wants to ask is "What if?" It's the thought that more could have been done to avert a tragedy. Countless losses, injuries, and deaths could surely be avoided if business owners and managers had better emergency preparations, precautions, and protocols in place. Too often, however, protective measures are low on the list of a company's priorities.

Hindsight is great. Too bad it comes so late. Thus, to avoid the avoidable, devote some time and money now to improve your company's protections and safeguards, because you don't want to be burdened someday by the thought "If only I'd . . ."

THE THREAT MATRIX

No company is immune from crime or terrorism. However, because terrorists seek publicity, large, well-known companies are more likely to become targets than are small, lesser-known ones. This doesn't mean, though, that simply because your company is small or relatively obscure that it's out of danger. Terrorists might, for instance, target the building your firm is housed in, confuse your company name with that of another, or pick your name and location at random. Your business might even suffer collateral damage from a terrorist attack next door. Thus, while larger firms require the highest level of protection, all businesses, regardless of size or notoriety, should know the basics of antiterrorism protection, take appropriate preventative measures, and institute emergency-response and contingency programs.

The threat matrix for American business follows the lines of location, type of firm, time-related factors, and hierarchy. Companies based in major U.S. cities are at higher risk than those located elsewhere. In addition, department stores, supermarkets, and shops located in large malls or in high-density areas of major cities are at greater risk than others. Note, however, that this could change. If security becomes tight in major cities, terrorists might switch to softer targets in the suburbs. Still, the types of businesses most likely to be attacked include transportation (including cargo shipping), financial institutions, infrastructure (especially, electric power, water, and communications), and other vital industries (notably, oil refineries and oil and gas storage depots). Firms whose names are synonymous with "America" could see their foreign operations struck. Because terrorists seek to kill and maim as many people as possible, peak times of business operation also play a part in the threat matrix. Think in terms of rush-hour transportation, winter or summer vacation resorts, and crowded holiday shopping periods. Hierarchy is another consideration. Companies that are symbolic of America are at high risk, as are businesses vital to the U.S. economy. So are firms in industries or sectors that have been attacked before, such as airlines and banks. Money, too, can be a factor, if terrorists think they can steal large sums from a company or its customers.

It's a mistake, however, to become a static thinker. Terrorists will sooner or later break the mold, so expect the unorthodox. The FBI, according to the Associated Press, has already begun planning for the unexpected by "looking over the horizon." A new FBI intelligence unit has thus far warned scuba-diving schools to be wary of suspiciously large equipment purchases or inquiries about potential targets (e.g., ports, dams, water supply or treatment systems, and riverside power plants). Similar warnings have gone out to businesses that supply other materials that terrorists might use, such as fertilizer and chemicals employed in bomb-making. The FBI also has warned banks, Jewish schools, and synagogues to be on guard.

In selecting targets, terrorists balance importance against opportunity. A tempting target might be skipped if security is too tight. Thus, the most vulnerable businesses are those deemed to be high-priority targets with the weakest, most penetrable security. This reinforces our conclusion that terrorism survival is largely a matter of choice. The businesses least likely to be successfully attacked will be those with the best defenses.

SECURITY DOCTRINE

Counterterrorism expenditures aren't worth the nickel unless a company adopts a strict philosophy of security. Half-measures don't work. Senior management must believe in security for an antiterrorism program to be effective.

Security starts at the top.

The most important security guard in a company is its chief executive officer, or CEO. If he's not committed to a philosophy of security, the entire company is at risk. The head of a firm not only controls the budget and guides hiring policy, but his thinking also pervades a company. If he believes in security, other managers will follow—and so will employees. But if he's dismissive of the terrorist threat and security enhancements, the company will be highly vulnerable to attack. CEOs should indeed take a cue from their Israeli counterparts, who take a hands-on approach to security planning and insist that employees are prepared for any emergency.

✔ The Office of Homeland Security, in conjunction with the Business Roundtable, has established a new communications network to alert CEOs immediately to a terrorist attack and enable them to talk with one another and government officials instantly. Dubbed CEO COM Link, CEOs must be members of the Business Roundtable and receive approval before they can participate in the program. For information, call the Business Roundtable at 202-872-1260 or go to www.brtable.org.

Make personnel safety and security your highest priority.

In setting priorities, view security as a pyramid, with the safety and welfare of your personnel at the top. Next comes building security and systems security, including information technology. Finally, make backup systems and contingency planning your base. This structured approach should help in deciding matters of resource allocation, commitment of funds, timing, and other corporate policymaking.

Don't let cost govern security planning.

Senior executives must be concerned with corporate finances, but too often we've been told by managers, "Just fix the mailroom; we'll get to the rest later." Would they say the same thing about computers or communications? ("Just put computers on the second floor; we'll get to the rest later." Or, "Just hook up the Internet at headquarters; we'll get to the branch offices later.") Of course not. But when it comes to security spending, many business executives suddenly become penny wise and pound foolish. Don't be one of them.

Hire a qualified corporate security officer.

Nothing beats having a trained and experienced professional in charge of corporate security—and that means more than just hiring a retired policeman. Establish an office of security and put it in the hands of a seasoned professional, well versed in counterterrorism. Consider candidates with backgrounds in government agencies such as the FBI or CIA, or military organizations such as the Defense Intelligence Agency. If your company has a large overseas presence, look for former employees of foreign intelligence services or counterterrorism military officers.

✔ Find someone with an understanding of information technology, because cyberspace threats are likely to grow in the future and you'll need a security chief who can speak the same language as your computer and Internet technicians.

Be sure your corporate security chief attends state and local terrorism-threat briefings.

States and localities are beginning to set up formal procedures for terrorist-related information sharing with businesses in the community. Within a month of the September 11 disaster, New York State, for example, opened an Office of Public Security (www.state.ny.us/security/), which holds closed-door meetings with corporate security chiefs to discuss terrorist threats and responses. Insist that your corporate security director participate in such forums.

Take advantage of the FBI's national-security awareness program.

Knowledge is a key to effective counterterrorism, so take advantage of the information supplied to companies by the FBI. The Awareness of National Security Issues and Response (ANSIR) Program is the FBI's national-security awareness program. It's the "public voice" of the FBI for espionage, counter-intelligence, counterterrorism, economic espionage, cyber- and physical-infrastructure protection, and all national security issues. The program is designed to provide unclassified national-security threat and warning information to U.S. corporate security directors and executives, law enforcement, and other government agencies. Information is disseminated nationwide via the ANSIR-Email and ANSIR-FAX networks. Interested U.S. corporate officers should provide their e-mail address, position, company name and address, as well as telephone and fax numbers, to the national ANSIR e-mail address at ansir@leo.gov. An ANSIR coordinator will then contact you.

Report any hint of trouble to the proper authorities.

The FBI and local law enforcement, among other government agencies, are keenly interested in information that could, when pieced together, reveal an impending terrorist attack. In that regard, you and your company can play an important role by reporting all suspicious activity, no matter how seemingly innocuous, to your local police department or the FBI. FBI field offices, located in 55 major cities throughout the United States and in San Juan, Puerto Rico, can be found at www.fbi.gov/contact/fo/fo.htm.

Tap knowledgeable outside security consultants.

In-house security resources can only go so far in protecting a company because their breadth of knowledge and experience is limited. Professional security consultants—such as my own company, Interfor, Inc. (www.inter forinc.com or 212-605-0375), which operates in the United States and around the world—supply a unique service by double-checking systems, probing for weaknesses, and offering solutions. Outside consultants with solid backgrounds in counterterrorism are particularly useful in that they have done this before; they've typically conducted security checks at hundreds of companies and are able to draw on this experience in advising other firms.

Apply risk-management techniques in counterterrorism policymaking.

"Risk management is a systematic and analytical process that weighs the likelihood that a threat will endanger an asset, individual, or function and identifies actions to reduce the risk and mitigate the consequences of an attack," states the U.S. General Accounting Office in its October 12, 2001, report "Homeland Security: Key Elements of a Risk Management Approach" (GAO-02-150T). "A good risk management approach," it continues, "includes the following three assessments: a threat, a vulnerability, and a criticality. After these assessments have been completed and evaluated, key steps can be taken to better prepare the United States against potential terrorist attacks. Threat assessments alone are insufficient to support the key judgments and decisions that must be made. However, along with vulnerability and criticality assessments, leaders and managers will make better decisions using this risk management approach. If the federal government were to apply this approach universally and if similar approaches were adopted by other segments of society, the United States could more effectively and efficiently prepare in-depth defenses against terrorist acts." For a copy of the report, go to www .gao.gov.

Institute a notification plan and chains of command.

Emergencies often get out of control because of a lack of timely information and confusion over who's in charge. To avoid these pitfalls, institute a notification program for the timely central collection of information and create clear chains of command. This combination should make for better decision-making in times of crisis. Be sure, too, that all bases are covered, and build redundancy into the system, so that if someone is unavailable, his responsibilities will be taken over automatically by another, preassigned company official.

✔ Equip key personnel with pagers and design "phone trees" (i.e., lists of names and telephone numbers for personnel to call in emergencies; by assigning a limited number of calls to each person, a large number of staff members can be reached in a short time) to send out alerts speedily.

Periodically review all security policies and procedures.

Terrorist targets and techniques evolve, so ensure that your security policies and procedures keep pace by conducting periodic reviews. Continuously assess the changing nature of the threat, look for new vulnerabilities at your facilities and in your operations, and conduct checks to verify that all existing security procedures are being followed correctly throughout your enterprise. You might even have teams of specialists probe for weaknesses.

PERSONNEL SAFETY

Never forget that your most important corporate assets are your people. Make their safety and security your first priority.

Educate your workforce in counterterrorism.

Vigilant employees are a company's most valuable counterterrorism tools. The sheer number of eyes and ears helps to reduce the dangers. So educate your labor force—from top to bottom—in antiterrorism. Provide seminars, workshops, training sessions, and brochures, explaining to employees what to look for and how to handle emergencies. Deploy suggestion boxes—and reward good ideas.

✔ Start with the little things to let employees know it's no longer business as usual. For instance, insist that the handwriting on all sign-in and sign-out sheets be legible.

Embed a culture of security within your company.

Use every opportunity to remind employees of the importance of security. The old World War II posters saying "Loose lips sink ships" actually worked by reminding military personnel, dockworkers, and civilian defense employees to keep secrets. So consider posters, brochures, training films, e-mail, focus groups, and the like. And have senior executives set an example by raising the issue of security at meetings and in everyday discussions.

✔ Instruct employees to treat door access codes and ID badges with the same care they give to protecting their credit cards and bank personal identification numbers (PINs).

Tell executives, in particular, to vary their daily routines.

Terrorists and other criminals often rely on a person's normal routines to help them carry out their attacks. Sticking with the same schedule and pattern every day lets attackers get their timing down pat. It gives them confidence so that they can lie in wait and target their victim with relative ease. Instruct personnel, particularly executives, to vary their routines. Leave home and work at different times. Drive different routes to work. Don't park in the same spot each day. And if they feel especially vulnerable, tell them to switch cars once in a while.

Remove VIP signs in company parking lots.

Parking lot signs with the titles or names of a company's senior management are dead giveaways to anyone meaning to do them harm. Indeed, signs of any sort indicating VIP parking pose the same problem. Someone seeking to plant a bomb, fatally disable a vehicle's brakes, or shoot or kidnap a high-level corporate executive can use those signs to select his or her target. Make your executives less vulnerable to attack by ridding your facilities of all parking space indicators.

Instruct employees in safe methods of business travel at home and abroad.

Formally instruct your employees in safe methods of business travel, following the recommendations laid out in this book. Don't let your employees feel as if the company abandons them the minute they leave on a trip. Assure them that their travel safety is a corporate concern. Supply employees headed overseas with copies of publications like "Personal Security Guidelines for the American Business Traveler Overseas." Published by the U.S. State Department's Overseas Security Advisory Council, the guide can be downloaded at www.ds.state.gov/about/publications/osac/personal.html.

✔ To protect proprietary information, have your business travelers and foreign-based employees read "Guidelines for Protecting U.S. Business Information Overseas," which outlines the steps to take to protect sensitive information when living, working, or traveling abroad. It's at www.ds.state.gov/about/publications/osac/protect.html.

PHYSICAL SECURITY

Security is determined by the barriers you build. The more sophisticated and extensive the barriers, the harder and longer it will take to get through them.

That delay could mean the difference between life and death in a terrorist incident. So think of security as a lock that buys precious time.

Survey your premises and operations for vulnerabilities and ease of escape.

Terrorists normally conduct surveillance of potential targets before settling on a specific location. They look for openings and weigh the odds of being caught. Most terrorists, apart from suicide bombers, consider escape a key criterion in selecting a target. So, do as the terrorists do. Survey your offices, stores, factories, buildings, and operations with an eye toward their vulnerability to attack, and determine how easy it would be for an attacker to evade capture. Then, institute changes to lessen your risk of attack and increase a terrorist's (or a common criminal's) risk of capture.

The Treasury Department's Bureau of Alcohol, Tobacco, and Firearms has published an extremely useful guide titled "Bomb Threats and Physical Security Planning" (available online at http://www.atf.treas.gov). Here are a few of its recommendations: Patrol potential hiding places (e.g., stairwells, restrooms, and vacant offices). Lock doors to boiler rooms, mailrooms, computer centers, laboratories, switchboards, and control rooms. Institute a procedure to account for keys. Keep Dumpsters under surveillance to prevent the planting of a bomb. Handle combustible materials carefully.

✔ Take similar precautions to prevent burglary and robbery. The Los Angeles Police Department offers some helpful advice under the heading "LAPD Crime Prevention Lesson Plans," which can be found by clicking "Crime Tips" at www.lapd.org.

🚫 Don't overlook heating, ventilation, and air-conditioning systems. Inspect them for ease of access and the potential for tampering, because biological, chemical, or radiological agents could be dispersed throughout a building or set of offices via air ducts. Similarly, recirculated air from a mailroom contaminated with, say, a biological agent could spread disease to other parts of the building.

Erect concentric rings of defense.

View your security system from the outside in and then erect concentric rings of defense, so that if an outer ring is penetrated, the attackers will be stopped by an inner one. These rings of defense should include physical barriers, such as doors, fencing, and lighting, as well as behavioral barriers, such as procedures, routines, and employee awareness.

Install video security and alarm systems.

A good way to determine whether your premises are under surveillance by terrorists or others is to use a closed-circuit video security system, which allows you to spot any repeated appearances of the same suspicious person or persons on or around your property. Video recordings are also helpful to law enforcement when trying to identify criminal suspects. Silent alarms, connected to your security center, should be deployed on exterior entryways and key interior doors.

 Store recorded videotapes for at least a month before reusing.

Place crash-proof barriers around building entries.

To prevent a car bomber from driving straight into a ground-floor lobby or other building entry, install crash-proof barriers. Be careful, however, of the type of barrier you select. For instance, large, concrete flower pots may look attractive, but they are also good places to conceal bombs. Unless such pots are watched 24 hours a day, seven days a week, they're a security liability. Buy solid concrete blocks instead.

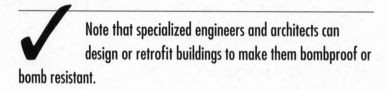 Note that specialized engineers and architects can design or retrofit buildings to make them bombproof or bomb resistant.

Maintain perimeter defenses around large facilities, and pay attention to landscaping.

Fencing, security gates, lighting, and even landscaping should all be part of your perimeter defenses around factories and other complexes. Check your property to see if it has places where intruders could easily hide or penetrate perimeter defenses undetected. Keep ground cover, such as hedges, cut low, and leave a gap of several yards between buildings and shrubbery. Also, keep a good distance between parking lots and building walls, for fear of car bombs. The Bureau of Alcohol, Tobacco, and Firearms recommends a minimum distance of 300 feet. If that's not possible, have security screen all nonemployee drivers and get approval for entry before these cars access the premises. The Royal Canadian Mounted Police's Technical Security Branch offers an excellent discussion of physical security and related matters in a variety of publications that are available online at www.rcmp-grc.gc.ca/tsb/pubs/index.htm.

 Enclose truck bays and maintain a security presence in shipment areas to prevent building penetration by intruders.

Reconsider your occupancy of space in a building with public underground parking.

Underground garages in office buildings, particularly in high-risk locales, may become prime terrorist targets. Terrorists would, of course, have to gain entry, so the most vulnerable underground garages are those that are open to the public. It is doubtful that garage attendants or even security personnel could stop terrorists from parking a vehicle loaded with explosions beneath a targeted building. Unless underground parking is restricted to the building's tenants and entry requires opening a locked garage door or the approval of a guard, the building simply isn't safe. If your business occupies space in a building with unrestricted public parking in an underground garage, you ought to consider moving, especially if you're located in a high-risk city or other location. The same holds if your office adjoins an outdoor public parking lot.

✔ In assessing your risk of terrorist attack, consider your business's proximity to any higher-risk targets in neighboring locations, because you could suffer collateral damage.

REDUCING VULNERABILITIES

"If you have reason to believe that you are likely to be a terrorist target because of the nature of your business," advises the British Home Office, "you should anticipate that terrorists will do research to work out where your greatest vulnerability is. What material about you is in the public domain? What published facts point to installations or services that are vital to the continuance of business? What might attract attention as a prestige target even though its loss may not mean immediate business collapse? Giving thought to what matters to you, and what is most vulnerable, will enable you to make realistic plans for deterring terrorist attack and minimising [*sic*] the damage should one occur at or near your premises" ("Bombs: Protecting People and Property," 4th edition, Home Office Communication Directorate, 1999).

Know who is in your offices at all times.

Electronic identification cards should be distributed to every employee at all but the smallest of companies and required to be worn at all times. They should be tied into a central computer to provide a real-time accounting of everyone in the building. The only way to accomplish this, though, is to close the loop. In other words, require that ID badges be swiped upon both entering and exiting the building. If there is no notification requirement upon egress, you'll never have a true account of who's actually in the building. The same holds for office temporaries, outside contractors, and guests.

 Consider biometric means of identification (e.g., facial or eye recognition systems) if circumstances warrant.

 Don't forget to maintain security at the building's freight entrance.

Restrict freedom of movement within offices and buildings.

Monitor internal movements within a building or office electronically through the use of surveillance cameras and mini-checkpoints, which needn't be manned but must be tied to your security office. Personnel should be required to swipe an ID card (or provide retinal or facial scans) to gain entrance to a floor. On large floors, additional gateways should be established to monitor movement from department to department. If highly sensitive work is conducted at a company, ID badges should limit employee access to those areas in which his physical presence is required. All other areas should be off-limits to nonessential personnel, unless special clearance is given—and that clearance should be time-restricted (i.e., good for, say, only an hour or a day at a time). The idea is to create zones within an office or building so as to limit unrestricted movement. The less freedom a terrorist (or a common criminal) has to move about a building or office, the less able he'll be to get to his target.

ID check and escort all visitors.

All visitors must be required to sign a logbook and show proper identification. Calls should be made to confirm the appointments before access is permitted. Every visitor should have an escort while in the building. At no time give a visitor a pass that would take him any further than the nearest washroom. Finally, require that each visitor sign out before leaving the building.

✔ **When querying first-time visitors to your facilities, have security ask questions that require full-sentence answers, not just a yes or a no. Such responses tend to reveal a visitor's bona fides.**

Get employees into the habit of questioning unfamiliar faces.

One of the most common means of getting past checkpoints, such as keypunch doors, is simply to walk in behind an authorized staff member once the door has been unlocked. Make it a requirement, therefore, for personnel to question anyone unfamiliar who is trying to enter a security door without using his own ID and password. (Again, station cameras at all of these locations so that a suspicious person can be spotted quickly and security personnel sent to the scene promptly.) Further instruct employees to question any unfamiliar person roaming around the offices without an escort.

Background-check the credentials of all contractors, including locksmiths.

Don't take anything for granted when allowing outside contractors onto your premises. Check their credentials thoroughly before hiring them. In January 2002, for example, U.S. officials told NBC News that five terrorist suspects in Bosnia were involved in a plot to infiltrate and blow up the U.S. embassy in Sarajevo. The job reportedly involved embassy insiders. NBC News was told that as part of the well-planned plot, one of the suspects had married the daughter of the Bosnian locksmith who worked for the U.S. embassy. Over time, the locksmith provided the suspects with keys, codes, and combinations to embassy locks and security devices.

Maintain a list of all persons fired for cause or who have made threats against the company or its employees.

Sadly, many workers and managers have been attacked by former employees and others seeking revenge. It's essential, therefore, to maintain a list of potential threats, including workers and contractors fired for cause and seriously disgruntled clients who have threatened violence. Don't hesitate to contact the police if you believe lives are in danger.

✔ Make sure former employees no longer have password access to your computer systems, keys to offices, or company cars. Be sure, too, that they've surrendered all corporate ID badges. Change locks and door codes if you have serious concerns.

Keep a central file of all hate and threatening mail.

Employees throughout your company should be instructed to forward any hate mail or mailed threats to the director of security, where the items should be tagged, bagged, and kept on file for future reference. Contact law enforcement or the FBI in serious cases.

Sterilize the information you provide about your company.

Terrorists often look for vulnerabilities by using the Internet and other publicly available sources to find information about companies, locations, key facilities, etc. Maps and photographs are particularly useful. Be sure you aren't playing into the terrorists' hands by disseminating information that could be used in target selection and the planning of an attack. Carefully scrutinize all of your published materials, and to the extent possible, sterilize that information by removing key details that could aid a potential attacker. Take an especially close look at the contents of your company's Web site.

Check the manifests of any shipments from overseas.

Because of the danger of bombs of various sorts being shipped via container by terrorists overseas, have the manifests for all international shipments sent in advance. Check the contents of shipping containers against the manifest. If a suspicious item is found, evacuate the area and notify security and law enforcement.

Institute counterterrorism programs at your foreign operations.

Terrorists are out to assault "America" and that includes U.S. business interests located overseas. So don't neglect to institute counterterrorism programs at all of your foreign operations. Maintain open lines of communications, and have your security chief serve as a liaison. Senior management should be notified of and approve all overseas security and contingency plans. Consult "Security Guidelines for American Enterprises Abroad," a compilation of security guidelines for American private-sector executives operating outside the United States. Published by the U.S. State Department's Overseas Security Advisory Council, it can be found at www.ds.state.gov/about/publications/osac/enterprises.html.

EMERGENCY PREPAREDNESS

Traditional emergency plans are insufficient to deal with acts of terror. Relying solely on fire drills just doesn't cut it anymore. Rethink your company's emergency preparations and training with a view toward such terrorist threats as bombings and biological attacks.

Develop building or office evacuation plans.

For employees to react sensibly in a crisis, they ought to have an evacuation plan already down pat. Plan for both complete and partial building evacuations. There should be designated routes of exit and a means of communicating with employees to ensure that they don't run smack into danger. Train marshals, equipped with walkie-talkies or cell phones, to direct the flow of people, and have an evacuation command center that's capable of mobility. (You don't want anyone staying inside a hazardous building.) Consult with local emergency services when devising the plan. For more discussion, see the British Home Office's "Bombs: Protecting People and Property—A Handbook for Managers" at www.homeoffice.gov.uk/atoz/terrorists.htm. Also, the Chicago Fire Department offers online the extremely useful "Suggested High Rise Office Building Evacuation Plan" at www.ci.chi.il.us/Fire/, explaining the roles of floor wardens, floor leaders, searchers, stairwell and elevator monitors, and aides to the handicapped.

 Two-way radio communications and cellular telephones should never be used if a bomb is found or suspected, for the transmissions could trigger a detonation.

Make provisions for evacuating the disabled or elderly.

The U.S. Architectural and Transportation Barriers Compliance Board (aka Access Board), a federal agency, has designed an emergency evacuation plan to meet the needs of the disabled, including training in the use of evacuation chairs. Go to www.access-board.gov to get a copy of the plan. For more information, contact the Access Board at Suite 1000, 1331 F Street NW, Washington, DC 20004-1111, 202-272-0043, 800-872-2253, or 202-272-0080 (TTY 202-272-0082).

Regularly conduct emergency drills and hold employee seminars.

In this age, emergency drills are essential for businesses. These should go beyond mere fire drills and include employee seminars, floor drills, and whole building evacuations. The British government offers a useful "Exercise Planners Guide" for "managers, executives, chief officers, and others who decide their organization's overall strategy for contingency planning, including training and exercising, to help prioritize the allocation of resources." The guide can be found at www.ukresilience.info/contingencies/cont_bus.htm.

As for seminars, the Oak Ridge National Laboratory insightfully notes in an instructor's manual on chemical warfare that adults "strongly resist learning anything merely because someone says they should" and "will learn only what they feel a need to learn." For tips on holding successful employee seminars, consult "Instructor Guide for Techniques for CSEPP [Chemical Stockpile Emergency Preparedness Program] Instructors," by Edith Jones of the Oak Ridge National Laboratory, which is available at http://emc.ornl.gov/emc/PublicationsMenu.html.

Have fire marshals review your emergency evacuation systems.

Whether or not you own the building or buildings in which your offices are housed, request a safety inspection from the local fire department. Stress your concern that evacuation routes, such as emergency stairs and fire doors, must be adequate to get your people out in a hurry. Also, have the fire department review your evacuation plans and training procedures, then make any recommended improvements.

Routinely inspect all fire doors and emergency staircases to ensure they aren't locked or blocked.

Some of the most tragic stories to emerge from the World Trade Center were those that told of fire doors that were locked. In some cases, the victims could do little more than call their loved ones to say a last good-bye. Don't let that happen at your company. Make it a matter of course to inspect all fire doors to ensure that they are never locked, and have your security personnel regularly transit every emergency route out of your building to make sure there are no bottlenecks or impassable barriers. If that means installing new security devices to prevent theft or trespass, make the investment. Never sacrifice safety just to save a few dollars.

Develop nonevacuation emergency plans to stay safely indoors.

In some instances, such as attacks with an aerosolized chemical or biological weapon or "dirty bombs" that release radioactive debris, fleeing a building into the contaminated streets would be unwise. Your company, therefore, should also develop plans to keep employees (and customers) within the safety of your building in such an event. The flow of air from an attack site near your building is your biggest worry. Turn off heating, ventilation, and air-conditioning systems, and close all windows and doors. Interior rooms, without any means for air from the outside to seep in, are good places for employees to huddle until rescuers arrive or they're told it's safe to leave the building.

✓ To create "safe rooms" into which lethal agents won't penetrate, store thick plastic sheeting and duct tape, and tell workers to fill any keyholes or other large gaps with paper or cloth before taping them over.

Stockpile emergency supplies and protective masks throughout your building or office.

In case egress from an office is impossible, emergency stockpiles of water and nonperishable food should be stored at several designated locations on every floor. A first-aid kit, cell phone, flashlights, whistles, portable radio, and extra batteries should also be stored at these locations. The storage site should *never* be locked. In addition, since infrastructure, notably power plants, are high on the terrorists' list of targets, prudence would dictate installing backup electrical generators. Furthermore, in case of a chemical-biological attack, have plenty of inexpensive N95 facial masks on hand to protect against the inhalation of deadly agents.

Not every company is going to provide employees with emergency equipment, so you could take the opportunity to give your customers pocket-size flashlights, embossed with your company's name, phone number, etc.

If you receive an unexpected flashlight, be certain you know the sender. The Bureau of Alcohol, Tobacco, and Firearms warns that flashlights are often rigged as bombs.

Adapt the Israeli concept of "protected space," especially at high-risk facilities.

The Israeli Ministry of Defense has long promulgated the concept of "protected space" or protective shelters among the civilian population and businesses. "The Protected Space is a handy, easy-to-reach space capable of providing those staying in it with protection against both conventional and nonconventional weapons for several hours," explains the Israel Home Front Command ("Protected Space/Shelters—Guidelines," Home Front Command, Israel Defense Forces). Since 1992, every new building or addition to an existing building in Israel has, by law, been equipped with either an Apartment Protected Space or a Floor Protected Space (FPS). Each FPS has a blast door that opens outward, a filtered ventilation system, and emergency lighting, plus a telephone and connections for TV and radio reception. The rooms are stacked one above the other, extending the full height of a building, so that they form one contiguous tower of secure rooms. Sealed openings in the rooms' floors and ceilings interconnect the FPSs to facilitate escape or rescue.

Details can be found at www.idf.il/english/organization/homefront/home front2.stm.

Follow safe mail handling techniques.

See chapter 4, "Mail Handling Guidelines," for the dos and don'ts about your company's mail.

BOMB THREATS AND EXPLOSIONS

The first thing to know about bombs is that they hardly ever look like bombs. Bombs can be made to resemble anything and often appear innocuous. For example, the bomb that exploded in a busy Hebrew University cafeteria in Jerusalem in July 2002, killing five Americans and two Israelis, was left in a bag on a table and was detonated by a cell phone. Next, err on the side of caution. When in doubt, alert the authorities. Finally, tell employees that receive a telephoned bomb threat to stay calm, get as much information as possible from the caller, use the bomb-threat checklist underneath their telephones, be attentive to any overheard background sounds or statements, and contact security or the police immediately after the call.

Keep an eye out for unattended objects and suspicious packages.

Employees should be alert to any packages, briefcases, luggage, handbags, and the like left unattended in the lobby of your building or near the outside of your building, especially in front of large glass windows. Flying glass can be deadly in an explosion. Pay particular attention if someone gingerly places, rather than drops, a package, bag, or other item in an unusual place or an out-of-sight spot, such as the back of a store shelf or behind plants in a building lobby. Be equally alert to any other suspicious behavior in or near your building. Look for signs of nervousness in the person. Watch to see if he appears to be on the lookout for police or security guards. Be leery of anyone trying— often repeatedly—to enter a locked door or an emergency door. Contact building security or the police, and provide them with a description of the person and the location of the suspicious object.

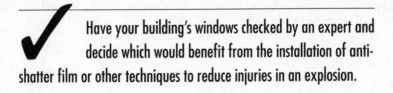

Have your building's windows checked by an expert and decide which would benefit from the installation of anti-shatter film or other techniques to reduce injuries in an explosion.

Take all bomb threats seriously and immediately report them to law enforcement.

Because most bombs are homemade, any caller to your company warning of an imminent explosion could very well be the bomb-maker himself and should be taken seriously. It should go without saying that bombs are nothing to be toyed with by amateurs. Let law enforcement and trained bomb-disposal experts handle things.

Put a bomb-threat checklist under every office telephone.

The Bureau of Alcohol, Tobacco, and Firearms has compiled the following checklist to use in the event of a telephoned bomb threat. The checklist can also be employed in handling biological or other terrorist threats. A copy of the checklist, contained in a protective paper sleeve or plastic bag, should be placed under every office telephone and at switchboards. All employees should be trained to follow these procedures, the aim being to get as much information from the caller as possible and to write down anything overheard. Finally, set up primary and secondary telephone numbers for them to call if they do receive a bomb threat.

Develop a bomb-incident plan.

We needn't be mere passive victims of bombings or even bomb threats. Advance preparations can alleviate panic and may save lives in the event of a blast. "If there is one point that cannot be overemphasized," says the Bureau of Alcohol, Tobacco, and Firearms (ATF), "it is the value of being prepared. Do not allow a bomb incident to catch you by surprise."

The big question is whether to evacuate immediately following a bomb threat. The ATF says that initiating a search after a threat is received is "per-

Bomb Threat Checklist

Exact time of call: _____

Exact words of caller:

QUESTIONS TO ASK

1. When is the bomb going to explode? _____

2. Where is the bomb?

   _____

3. What does it look like?

   _____

4. What kind of bomb is it?

   _____

5. What will cause it to explode? _____

6. Did you place the bomb?

   _____

7. Why?

   _____

8. Where are you calling from?

   _____

9. What is your address?

   _____

10. What is your name?

    _____

CALLER'S VOICE (circle)

Calm	Slow	Crying	Slurred
Stutter	Deep	Loud	Broken
Giggling	Accent	Angry	Rapid
Stressed	Nasal	Lisp	Excited
Disguised	Sincere	Squeaky	Normal

If voice is familiar, whom did it sound like? _____

haps the most desired approach," because it's "not as disruptive as an immediate evacuation." We, however, recommend this decision be made at the highest level of your company. Have your crisis management team and security officer weigh the pros and cons of immediate evacuation and make recommendations to senior management, on whose shoulders the final decision should rest.

In planning for bomb threats, the ATF recommends the following steps:

Bomb Incident Plan

(1) Designate a chain of command.

(2) Establish a command center.

(3) Decide what primary and alternative communications will be used.

(4) Establish clearly how and by whom a bomb threat will be evaluated.

(5) Decide what procedures will be followed when a bomb threat is received or device discovered.

(6) Determine to what extent the available bomb squad will assist and at what point the squad will respond.

(7) Provide an evacuation plan with enough flexibility to avoid a suspected danger area.

(8) Designate search teams.

(9) Designate areas to be searched.

(10) Establish techniques to be utilized during search.

(11) Establish a procedure to report and track progress of the search and a method to lead qualified bomb technicians to a suspicious package.

(12) Have a contingency plan available if a bomb should go off.

(13) Establish a simple-to-follow procedure for the person receiving the bomb threat.

(14) Review your physical security plan in conjunction with the development of your bomb incident plan.

Command Center

(1) Designate a primary location and an alternative location.

(2) Assign personnel and designate decision-making authority.

(3) Establish a method for tracking search teams.

(4) Maintain a list of likely target areas.

(5) Keep floor diagrams in the center.

(6) Establish primary and secondary methods of communication. Caution: The use of two-way radios during a search could cause premature detonation of an electric blasting cap.

(7) Formulate a plan for establishing a command center if a threat is received after normal business hours.

(8) Maintain a roster of all necessary telephone numbers.

For a more detailed discussion, see "Bomb Threats and Physical Security Planning," a pamphlet published by the ATF, which is available online at http://www.atf.treas.gov or through the U.S. Government Printing Office.

Disseminate guidance on what employees should do in a building explosion.

Distribute information to employees on what to do in the event of a building explosion or an attack with chemical or biological weapons. Reprinted here verbatim is guidance from the Federal Emergency Management Agency (www.fema.gov/hazards/terrorism/terrorf.shtm):

DURING: In a building explosion, get out of the building as quickly and calmly as possible. If items are falling off of bookshelves or from the ceiling, get under a sturdy table or desk. If there is a fire:

- Stay low to the floor and exit the building as quickly as possible.
- Cover nose and mouth with a wet cloth.
- When approaching a closed door, use the palm of your hand and forearm to feel the lower, middle, and upper parts of the door. If it is not hot, brace yourself against the door and open it slowly. If it is hot to the touch, do not open the door—seek an alternate escape route.
- Heavy smoke and poisonous gases collect first along the ceiling. Stay below the smoke at all times.

AFTER: If you are trapped in debris:

- Use a flashlight.
- Stay in your area so that you don't kick up dust. Cover your mouth with a handkerchief or clothing.
- Tap on a pipe or wall so that rescuers can hear where you are. Use a whistle if one is available. Shout only as a last resort—shouting can cause a person to inhale dangerous amounts of dust.

Assisting Victims: Untrained persons should not attempt to rescue people who are inside a collapsed building. Wait for emergency personnel to arrive.

Follow CDC guidance in attacks with chemical or biological weapons. See pages 85–89.

CONTINGENCY AND CONTINUITY PLANNING

The British government warns that "around half of all businesses experiencing a disaster and which have no effective plans for recovery fail within the following 12 months" ("How Resilient Is Your Business to Disaster?" UK Resilience, British Civil Contingencies Secretariat, 2002).

If you plan ahead for terrorist attacks or other crises and if your employees already know what to do when all goes wrong, your ability to limit the damage—in terms of lives, injures, and costs—and to maintain postcrisis business continuity is greatly improved.

Recognize that contingency and continuity planning works.

Preplanning can make the difference between the survival and demise of your business following a terrorist incident or other disaster. "A few years ago," says a U.K. government handbook, "a terrorist bomb seriously damaged the headquarters of a large insurance company over a spring weekend. By

Monday morning, furniture, computers, telephones, and supplies had been delivered to a relocation address and over 500 staff were at work. This could not have been done without careful planning, which had been tested by exercising, and as a result jobs were preserved and the business continued to flourish." See "How Resilient Is Your Business to Disaster?" at www.ukresilience .info/contingencies/cont_bus.htm.

Develop IT contingency plans.

A business's dependence on IT opens it up to a variety of disruptions, such as power outages, equipment failures, and terrorist attacks. IT contingency plans are, therefore, necessary. The National Institute of Standards and Technology, in June 2002, issued an extremely useful guide, covering everything from desktop and portable computers to local area networks and mainframe systems. In "Contingency Planning Guide for Information Technology Systems," it recommends that seven progressive steps be taken: (1) develop a contingency planning statement; (2) conduct a business impact analysis; (3) identify preventive controls; (4) develop recovery strategies; (5) develop an IT contingency plan; (6) plan testing, training, and exercises; and (7) plan maintenance. The guide is available at http://csrc.nist.gov/publications/nistpubs/ index.html.

Preplan a crisis communications strategy.

To address the concerns of customers, investors, employees, and the general public following an attack affecting your company or industry, have a crisis communications strategy in place ahead of time. We suggest taking a look at "The Business Roundtable's Post-9/11 Crisis Communications: Best Practices for Crisis Planning, Prevention and Continuous Improvement," published in June 2002 and available at www.brtable.org. It offers an array of useful advice. "Several Best Practices were reinforced or have emerged due to the events of 9/11 and continuing concerns about terrorism," it says, citing seven steps recommended by communications experts. These are (1) establish a full-time commitment to crisis management; (2) employ communications techniques to maximize crisis prevention; (3) designate backups; (4) keep vendors and consumers in the loop; (5) address terrorism as a global concern;

(6) be sensitive to the communication of risk; and (7) understand the pros and cons of the Web. To which we would add (8) keep investors and financial analysts in the loop, too.

Find a safe haven for duplicate business records, contracts, and the like.

Keep duplicates of important business records, contracts, patents, and copyrights at a safe, off-site location. Your lawyer's office is the most likely place. Otherwise, a bank safe-deposit box is always an option.

Store a duplicate directory of employee information at a secure, off-site location.

A duplicate employee directory stored safely at an off-site location could prove essential if you are denied access to your office for security reasons or in the event your office is damaged or destroyed in a terrorist attack. Larger companies could easily store vital employee information electronically at more than one site. A simple and safe approach for small companies would be to save duplicate personnel information on computer disks and then store them—along with printed backup copies—at the homes of senior management or similarly secure sites (e.g., a safe-deposit box or a lawyer's office). Keep these records up-to-date. Similarly, it could prove helpful (especially to law enforcement) to maintain at an off-site location the personnel records of former employees. If your organization teaches students, store their vital statistics at an off-site location and keep the information current.

Maintain a current list of employee emergency contacts.

Require all employees to fill out forms listing the name, address, and telephone numbers (both home and office phone numbers) of the person or persons they wished contacted in an emergency—and periodically send out reminders to employees to keep their contact information current.

Have a 1-800 number and a Web site that employees can consult for information.

Offer employees an option to submit DNA samples and fingerprints.

The tragedy of the World Trade Center attack showed how useful DNA samples and fingerprints can be in identifying remains. Inform your employees that they can voluntarily store samples of their DNA and fingerprints. A caveat, however, is in order: medical science is showing an ability to discover predisposition to disease and illness in a person's DNA. Some employees may treat an offer to store DNA samples skeptically, thinking (wrongly or rightly) that their company will test the samples for predisposition to illness and then, perhaps, terminate the worker or deny promotion. To put employees' minds at rest, include a written guarantee that the DNA information would only be used by the proper authorities for identification if the employee dies.

Plan ahead to maintain business continuity following a crisis.

Keeping your business running after a crisis can be difficult. To avoid large losses of revenue and to keep disruptions to a minimum, have a plan in place to maintain continuity. Appoint a business continuity planner as part of your crisis management team. Identify mission-critical operations and key personnel. Look for backup support that already exists at different locations. Maintain a contingency plan that includes emergency office space, equipped with computers and telephone lines; backup IT systems; emergency phone numbers; allocation of responsibilities to other offices or personnel; and preassigned emergency coordinators. Then, keep the plan current, updating it regularly to reflect changes in your business operations and resources.

The New York State Forum for Information Resource Management offers the online "Business Continuity Health Check" at www.nysfirm.org.

Get your board of directors' advice on your business continuity plans.

Business continuity management "is different from disaster recovery planning since it is proactive and concentrates on everything that is needed to continue the key business processes, whatever the catastrophe," notes a guide provided online by the British Department of Trade and Industry. Titled "Business Continuity Management—Preventing Chaos in a Crisis," written by Visor Consultants Ltd. of London, it lists eight steps to take to ensure postcrisis continuity. "In many organisations [*sic*] the concept of effective and routine Risk Management in parallel with regular Business Continuity Management, should crisis threaten, is being introduced and managed for the first time," it concludes. "However, in the final analysis it is the board of directors who must effect policies to ensure the company is acting responsibly and is suitably prepared to deal with crises." The guide is at www.dti.gov.uk/mbp/bpgt/m9ba91001/m9ba910011.html.

FINAL THOUGHTS

Call me old-fashioned, but I believe that it's better to be safe than sorry. I've never heard of any business overdoing its emergency planning and preparation. Keep in mind that criminals and terrorists aim to victimize you by finding the chinks in your armor. They are trying to outwit you by getting around the safeguards and protections that you already have in place. It's essential, then, to make anticrime and counterterrorism planning ongoing, including regular reviews and upgrades. And don't let cost be the deciding factor when it comes to investing in new safety measures or equipment. Many things are much more important than money.

✦ APPENDIX ✦

Where to Get Help and Information

T he following is a selection of sources for assistance and information relating to matters discussed in this book. The listings are grouped according to subject and are intended for reference purposes only. They should not be construed as an endorsement of a product or service, and they are not a guarantee of quality. Consumers are responsible for their own purchasing decisions and should exercise care. No liability shall attach to the author or publisher of this book. Note: In most cases, Web addresses have been abbreviated, leaving out the standard *http://* prefix.

Air Ambulances, Medical Evacuations, and Medical Escorts

U.S.-BASED INTERNATIONAL SERVICES:

Many of the U.S.-based services below accept collect calls from overseas, and many of the toll-free telephone numbers work anywhere in North America and the Caribbean.

Advanced Air Ambulance, Miami, Fla.; 800-633-3590, 305-232-7700, or 305-599-1100; fax 305-232-7734; www.flyambu.com.
Aero National, Inc., Washington, Pa.; 800-245-9987 or 412-228-8000; www.aeronational.com.
Air Ambulance America, Austin, Tex.; 800-222-3564 or 512-479-8000; www.airambulance.com.
Air Ambulance Network, Tarpon Springs, Fla.; 800-327-1966 or 727-934-3999; fax 727-937-0276; www.airambulancenetwork.com.
Air Ambulance Professionals, Fort Lauderdale, Fla.; 800-752-4195 or 954-491-0555; www.airambulanceprof.com.

AirEvac, Phoenix, Ariz.; 800-421-6111 or 800-321-9522; www.airevac.com.

Air Response, Inc., Denver, Colo., Clearwater, Fla., and Scotia, N.Y.; 800-631-6565 or 303-858-9967; fax 888-631-6565 or 303-858-9968; www.air response.net.

American Jet International, Houston, Tex.; 888-435-9254 or 713-641-9700; www.iflyaji.com.

Care Flight International, Inc., Clearwater, Fla.; 800-282-6878 or 727-530-7972; www.careflight.com.

Critical Care Medflight, Lawrenceville, Ga.; 800-426-6557 or 770-513-9148; fax: 770-513-0249; www.criticalcaremedflight.com.

Global Care, Inc., Alpharetta, Ga.; 800-860-1111; www.globalems.com.

International Association for Medical Assistance to Travellers, 417 Center Street, Lewiston, NY 14092; 716-754-4883; or 40 Regal Road, Guelph, Ontario, Canada N1K 1B5; 519-836-0102; www.iamat.org.

International SOS Assistance, Philadelphia, Pa.; 800-523-8930, 215-244-1500, or 215-245 4707; www.internationalsos.com.

Med Escort International, Inc., Allentown, Pa.; 800-255-7182 or 610-791-3111; fax 610-791-9189; www.medescort.com.

Medex Assistance Corp., Timonium, Md.; 888-MEDEX-00, 800-537-2029, or 410-453-6300; fax 410-453-6301; www.medexassist.com.

Medjet International, Inc., Birmingham, Ala.; 800-356-2161 or 205-592-4460; www.medjet.com.

Medway Air Ambulance, Lawrenceville, Ga.; 800-233-0655 or 770-934-2080; www .medwayair.com.

Mercy Medical Airlift (a charitable organization), Manassas, Va.; 800-296-1217 or 703-361-1191; www.patienttravel.org.

National Air Ambulance, Fort Lauderdale, Fla.; 800-327-3710 or 305-525-5538; www.nationaljets.com.

Travel Care International, Inc., Eagle River, Wis.; 800-524-7633 or 715-479-8881; www.travel-care.com.

Worldwide Assistance Services, Inc. (a unit of Paris-based Europ Assistance), Washington, D.C.; 800-777-8710 or 703-204-1897; www.worldwideassistance.com.

Foreign-Based International Services:

Austria: Austrian Air Ambulance, Vienna; 43-1-40-144; fax 43-1-40-155; www.oafa.com. Tyrol Air Ambulance, Innsbruck; 43-512-22-422; www.taa.at.

Canada: SkyService, Toronto/Montreal; 800-463-3482 or 514-497-7000; www.sky service.com.

China: Medex Assistance Corp. (U.S.-based), Beijing; 86-10-6465-1264; fax 86-10-6465-1269; www.medexassist.com.

Finland: Euro-Flite Ltd., Helsinki/Vantaa; 358-9-870-2544; fax 358-9-870-2507; www.jetflite.fi.

France: Medic Air, Paris; 33-1-41-72-1414; http://medicair.starnet.fr/.

Germany: German Air Rescue (DRF), Filderstadt; 49-711-70-10-70; www.drf.de.

Singapore: International SOS Pte Ltd., Singapore; 65-338-2311 or 65-338-7800; www.internationalsos.com.

South Africa: Medex Assistance Corp. (U.S.-based), Cape Town; 021-726-351; fax 021-751-478; www.medexassist.com.

Thailand: Siam Land Flying Co. Ltd., Bangkok; 011-662-535-6784; fax 011-662-535-4355; www.executivewings.com.

United Kingdom: International SOS Assistance (UK) Ltd., London, England; 44-0-20-8762-8000 or 44-0-20-8762-8008; www.internationalsos.com. Medex Assistance Corp. (U.S.-based), Brighton, England; 44-1273-22-3002; fax 44-1273-22-3003; www.medexassist.com.

Airline Links and Phone Numbers

Federal Aviation Administration; www.faa.gov/airlineinfo.htm.

Airport Links—U.S. and Foreign

Federal Aviation Administration; www.faa.gov/airportinfo.htm.

Air Travel—Consumer Complaints, Protection, and Reports

Aviation Consumer Protection Division, U.S. Department of Transportation, Room 4107, C-75, Washington, DC 20590; 202-366-2220 (TTY 202-366-0511); www.dot.gov/airconsumer/index.htm.

Federal Aviation Administration, 800 Independence Avenue SW, Washington, DC 20591; 800-255-1111; www.faa.gov.

Office of Aviation Enforcement and Proceedings, U.S. Department of Transportation, 400 Seventh Street, SW, Room 4107, Washington, DC 20590; www.dot.gov/airconsumer/index1.htm.

Air Travel—Flight Delays

Air Traffic Control System Command Center, Federal Aviation Administration; www.fly.faa.gov.

Air Travel Safety and Security

Aviation Safety Alliance, 1301 Pennsylvania Avenue NW, Suite 1100, Washington, DC 20004; 202-626-4104; www.aviationsafetyalliance.org.

Federal Aviation Administration, Consumer Safety and Security Complaints, 800 Independence Avenue SW, Washington, DC 20591; 800-255-1111; www.faa.gov.

Federal Aviation Administration, Flight Standards Service, International Aviation Safety Assessment Program (IASA); www.faa.gov/apa/iasa.htm.

Federal Aviation Administration, Office of Civil Aviation Security, Criminal Acts Against Civil Aviation; http://cas.faa.gov/crimacts/iacs.html.

National Transportation Safety Board, 490 L'Enfant Plaza SW, Washington, DC 20594; 202-314-6000; www.ntsb.gov/aviation/aviation.htm.

Air Travel Safety—Foreign Airlines

Aviation Safety Assessment Program, Federal Aviation Administration; 800-322-7873 or 800-255-1111; www.faa.gov/apa/iasa.htm.

Air Travel Security and Baggage Guidelines

Federal Aviation Administration, Information Hotline (normal business hours); 800-322-7873; www.faa.gov.

Transportation Security Administration, 400 Seventh Street SW, Washington, DC 20590; 866-289-9673; www.tsa.gov.

Air Travel Statistics

U.S. Department of Transportation, Bureau of Transportation Statistics; www.bts.gov.

Air Travel Warnings—United States and Abroad

U.S. Department of State, Travel Hotline; 202-647-5225.

U.S. Department of Transportation, Travel Advisory Line; 800-221-0673.

Air Travel—Weather Information

National Oceanic and Atmospheric Administration, National Weather Service; www.nws.noaa.gov.

Arson

Bureau of Alcohol, Tobacco, and Firearms, Arson & Explosives Division, 800 K Street NW, Room 680, Washington, DC 20001; 800-461-8841 or 202-927-7930; www.atf.treas.gov.

U.S. Fire Administration, 16825 S. Seton Avenue, Emmitsburg, MD 21727; 301-447-1000; fax 301-447-1052; www.usfa.fema.gov.

Bank Deposit Insurance Check

Federal Deposit Insurance Corp., 550 17th Street NW, Washington, DC 20429; 877-275-3342 or 202-942-3147; www2.fdic.gov/edie/.

Bank Fraud

Federal Deposit Insurance Corp., 550 17th Street NW, Washington, DC 20429-9990; 800-759-6596 or 877-275-3342; www.fdic.gov.

Federal Reserve Board, 20th Street and Constitution Avenue NW, Washington, DC 20551; 202-452-3000; www.federalreserve.gov.

Bioterrorism and Chemical Warfare

American College of Emergency Physicians, 1125 Executive Circle, Irving, TX 75038-2522; 800-798-1822; www.acep.org.

American Council on Science & Health, 1995 Broadway, Second Floor, New York, NY 10023-5860; 212-362-7044; www.acsh.org.

Center for Nonproliferation Studies, Monterey Institute of International Studies, 460 Pierce Street, Monterey, CA 93940; 831-647-4154; http://cns.miis.edu.

Centers for Disease Control and Prevention, Emergency Response Hotline (24 hours); 770-488-7100; 1600 Clifton Road, Atlanta, GA 30333; general information: 800-311-3435, 888-246-2675, or 404-639-0385; www.bt.cdc.gov.

Federal Emergency Management Agency, 500 C Street SW, Washington, DC 20472; 202-566-1600; www.fema.gov.

Federation of American Scientists, 1717 K Street NW, Suite 209, Washington, DC 20036; 202-546-3300; www.fas.org.

Food and Drug Administration, 5600 Fishers Lane, Rockville, MD 20857; 888-463-6332; www.fda.gov.

Food Contamination Hotline, U.S. Department of Agriculture; 800-535-4555; www.usda.gov.

Israeli Defense Forces, Home Front Command; www.idf.il/english/organization/homefront/index.stm.

Johns Hopkins University, Center for Civilian Biodefense Strategies; 410-223-1667; www.hopkins-biodefense.org.

MedlinePlus, U.S. National Library of Medicine; www.nlm.nih.gov/medlineplus/biologicalandchemicalweapons.html.

Oak Ridge National Laboratory, P.O. Box 2008, Oak Ridge, TN 37831; 865-574-4160; http://emc.ornl.gov/emc/PublicationsMenu.html.

Public Health Foundation, 1220 L Street NW, Suite 350, Washington, DC 20005; 202-898-5600; www.phf.org.

State Public Health Agencies Directory and Hotlines; www.statepublichealth.org.

U.S. Environmental Protection Agency, Environmental Emergency Hotline: 800-424-8802; 1200 Pennsylvania Avenue NW, Washington, DC 20460; 215-814-5000; www.epa.gov.

Virtual Naval Hospital, U.S. Navy Bureau of Medicine and Surgery and the University of Iowa; www.vnh.org.

Bombs

Bureau of Alcohol, Tobacco, and Firearms, Arson & Explosives Division, 800 K Street NW, Room 680, Washington, DC 20001; 800-461-8841 or 202-927-7930; www.atf.treas.gov.

Federal Bureau of Investigation, J. Edgar Hoover Building, 935 Pennsylvania Avenue NW, Washington, DC 20535-0001; 202-324-3000; www.fbi.gov.

British Terrorism-Related Web Sites

Defense Ministry; www.operations.mod.uk.
Emergency Newsroom; www.ukonline.gov.uk.
Foreign & Commonwealth Office; www.fco.gov.uk.
Foreign Travel Advisories; www.fco.gov.uk/travel.
News Coordination Center; www.ukresilience.info.
Prime Minister's Office; www.pm.gov.uk.
Scotland Yard, Metropolitan Police; www.met.police.uk.

Building Security

Bureau of Alcohol, Tobacco, and Firearms, Arson & Explosives Division, 800 K Street NW, Room 680, Washington, DC 20001; 800-461-8841 or 202-927-7930; www.atf.treas.gov.

Royal Canadian Mounted Police, Technical Security Branch; 613-991-9497; www.rcmp-grc.gc.ca/tsb/.

Cell Phones—International

Cellhire USA LLC, 45 Broadway, 20th Floor, New York, NY 10006; 866-CH-ONLINE; fax 212-376-7383; www.cellhire.com.

InTouch USA; 800-872-7626; international: 1-703-222-7161; fax 1-703-222-9597; www.intouchusa.com.

Japan Cell Phone Rentals, JCR Corporation, P.O. Box 15915, Honolulu, HI 96830; 800-611-7374, 808-924-5339, or 808-922-8655; www.jcrcorp.com.

Planetfone, 101 Convention Center Drive, Suite 700, Las Vegas, NV 89109; 888-988-4777; fax 888-388-4800; www.planetfone.com.

Rent Cell, 2625 Piedmont Road, Suite 56-170, Atlanta, GA 30324; 800-404-3093 or 404-467-4508; fax 810-454-1990; www.rentcell.com.

TravelCell; 877-235-5746; www.travelcell.com.

Travelers Telecom, 17141 Ventura Boulevard, Suite 204, Encino, CA 91316; 800-736-8123 or 818-325-2820; fax 818-325-2828; www.travtel.com.

VoiceStream; 800-937-8997 or 505-998-3793; fax 800-998-3666; www.voicestream.com.

CEO Security Services

Awareness of National Security Issues and Response, Federal Bureau of Investigation; e-mail: ansir@leo.gov.

CEO COM Link, Business Roundtable/Office of Homeland Security; 202-872-1260; www.brtable.org.

Interfor, Inc., Corporate Security, World Headquarters, 575 Madison Avenue, Suite 1006, New York, NY 10022; 212-605-0375; www.interforinc.com.

Charity Information and Verification

American Institute of Philanthropy, 4905 Del Ray Avenue, Suite 300, Bethesda, MD 20814; 301-913-5200; www.charitywatch.org.

BBB Wise Giving Alliance, 4200 Wilson Boulevard, Suite 800, Arlington, VA 22203; 703-276-0100; www.give.org.

Philanthropic Advisory Service, Council of Better Business Bureaus, 4200 Wilson Boulevard, Suite 800, Arlington, VA 22203-1838; 703-276-0100; www.bbb.org.

Check Theft/Fraud

CrossCheck; 707-586-0551; www.cross-check.com.

Equifax Check Systems; 800-437-5120; www.equifax.com.

International Check Services; 800-526-5380; www.intlcheck.com.

National Check Fraud Service; 843-571-2143; www.ckfraud.org.

National Processing Company; 800-255-1157; www.npc.net.

SCAN (Shared Check Authorization Network); 800-262-7771; www.scanassist.com.

TeleCheck; 800-710-9898 or 800-927-0188; www.telecheck.com.

U.S. Postal Inspection Service; 800-ASK-USPS; www.usps.com/postalinspectors/ welcome.htm.

Coping with Terrorism—Adults and Children

American Academy of Child and Adolescent Psychiatry, 3615 Wisconsin Avenue NW, Washington, DC 20016-3007; 202-966-7300; www.aacap.org.

American Medical Association, 515 N. State Street, Chicago, IL 60610; 312-464-5000; www.ama-assn.org.

American Psychiatric Association, 1400 K Street NW, Washington, DC 20005; 888-357-7924; www.psych.org.

National Center for PTSD (post-traumatic stress disorder); 802-296-6300; www.neptsd.org.

National Institute of Mental Health, 6001 Executive Boulevard, Room 8184, MSC 9663, Bethesda, MD 20892-9663; 301-443-4513; www.nimh.nih.gov.

U.S. Department of Transportation, Aviation Consumer Protection Division, Room 4107, C-75, Washington, DC 20590; 24-hour complaint line: 202-366-2220 (TTY 202-366-0511); www.dot.gov/airconsumer/.

U.S. Veterans Administration, Mental Health and Behavioral Sciences Services, 810 Vermont Avenue NW, Room 915, Washington, DC 20410; 800-827-1000; www.va.gov.

Credit Card Fraud, Loss, or Theft

Equifax, P.O. Box 740241, Atlanta, GA 30374; 800-525-6285; www.equifax.com.

Federal Trade Commission, Consumer Response Center, 600 Pennsylvania Avenue NW, Washington, DC 20580; 877-382-4357; www.ftc.gov.

MasterCard; 800-MC-ASSIST or 636-722-7111 (international collect calls accepted); www.mastercard.com.

TransUnion, Fraud Victim Assistance Division, P.O. Box 6790, Fullerton, CA 92634; 800-680-7289; www.tuc.com.

U.S. Postal Inspection Service; 800-ASK-USPS; www.usps.com/postalinspectors/ welcome.htm.

Visa Card; 800-847-2911 or 410-581-9994 (international collect calls accepted); http://usa.visa.com.

Credit Cards (Preapproved)—Opt Out

Main opt-out number: 888-5OPTOUT (or 888-567-8688).
Equifax; 800-525-6285; www.equifax.com.
Experian; 888-397-3742; www.experian.com.
TransUnion; 800-680-7289; www.transunion.com.

Credit Reports

Equifax, P.O. Box 740241, Atlanta, GA 30374; 800-685-1111; www.equifax.com.
Experian, P.O. Box 949, Allen, TX 75013-0949; 888-397-3742; www.experian
.com.
TransUnion, 760 Sproul Road, P.O. Box 390, Springfield, PA 19064-0390; 800-
916-8800; www.tuc.com.

Crisis Assistance—Families of U.S. Citizens Overseas

U.S. Department of State, Office of American Citizens Services and Crisis Management: 202-647-5225; or U.S. Department of State Operations Center Task Force:
202-647-0900; http://travel.state.gov/crisismg.html.

Cyberterrorism and Computer/Internet/Information Security

CERT (Computer Emergency Response Team) Coordination Center, Software Engineering Institute, Carnegie Mellon University, Pittsburgh, PA 15213-3890; 412-
268-7090 (24-hour hot line); www.cert.org.
Computer Incident Advisory Capability, U.S. Department of Energy; 925-422-
8193; www.ciac.org/ciac/.
Computer Security Resource Center, National Institute of Standards and Technology, 100 Bureau Drive, Mail Stop 8930, Gaithersburg, MD 20899-8930; 301-
975-2934 or 301-975-6478; http://csrc.nist.gov/index.html.
Forum of Incident Response and Security Teams, First.Org, Inc., PMB 349, 650
Castro Street, Suite 120, Mountain View, CA 94041; www.first.org.
National Infrastructure Protection Center, J. Edgar Hoover Building, 935 Pennsylva-
nia Avenue NW, Washington, DC 20535-0001; 202-323-3205; www.nipc.gov.
National Institute of Standards and Technology, Vulnerability and Threat Portal, 100
Bureau Drive, Stop 3460, Gaithersburg, MD 20899-3460; 301-975-NIST/6478;
http://icat.nist.gov/vt_portal.cfm.
New York State Office for Technology, Information Security Department, State
Capitol, Empire State Plaza, P.O. Box 2062, Albany, NY 12220; 518-474-0865 or
518-473-2658; www.oft.state.ny.us/security/security.htm.

Overseas Security Advisory Council, Cyber Threat Analysis and News; www.ds-osac
.org/edb/cyber/default.cfm.
Symantec Corp. (Norton products), 20330 Stevens Creek Boulevard, Cupertino, CA
95014; 408-517-8000; www.symantec.com (free online computer security check
and virus scan for home users).
U.S. Department of Justice, Computer Crime and Intellectual Property Section,
Criminal Division, 10th and Constitution Avenue NW, John C. Keeney Building, Suite 600, Washington, DC 20530; 202-514-1026; www.cybercrime.gov.

Deaths Abroad—U.S. Citizens

U.S. Department of State, Office of American Citizens Services and Crisis Management; 202-647-5225; http://travel.state.gov/crisismg.html.

Disaster Relief

American Red Cross, 430 17th Street NW, Washington, DC 20006; 877-272-7337;
www.redcross.org.
Federal Emergency Management Agency, 500 C Street SW, Washington, DC 20472;
800-462-9029; www.fema.gov.

Elderly and Disabled Assistance

AARP (formerly the American Association of Retired Persons), 601 E Street NW,
Washington, DC 20049; 800-424-3410; www.aarp.org.
American Red Cross, 430 17th Street NW, Washington, DC 20006; 877-272-7337;
www.redcross.org.
Federal Emergency Management Agency, 500 C Street SW, Washington, DC
20472; 800-462-9029; www.fema.gov.
Transportation Security Administration, 400 Seventh Street SW, Washington, DC
20590; 866-289-9673; www.tsa.gov.
U.S. Architectural and Transportation Barriers Compliance Board (aka Access
Board), Suite 1000, 1331 F Street NW, Washington, DC 20004-1111; 800-872-
2253 or 202-272-0080 (TTY 800-993-2822 or 202-272-0082); www
.access-board.gov.
U.S. Fire Administration, Federal Emergency Management Agency; www.usfa
.fema.gov.

Fire Safety

Chicago Fire Department, Public Education Section, 1010 S. Clinton Street, Chicago, IL 60607; 312-747-6691/92; www.ci.chi.il.us/Fire/.

Hotel-Motel Fire Safety Database, U.S. Fire Administration, Federal Emergency Management Agency; www.usfa.fema.gov/hotel/search.cfm.

Los Angeles Fire Department, 200 N. Main Street, Los Angeles, CA 90012; 213-485-5971; www.lafd.org.

New York City Fire Department, Office of Fire Safety Education; 718-999-2343/44; www.nyc.gov/html/fdny/html/safety/firesafety.html.

U.S. Fire Administration, 16825 S. Seton Avenue, Emmitsburg, MD 21727; 301-447-1000; www.usfa.fema.gov.

First Aid

American Association of Poison Control Centers, Poisoning Emergencies: 800-222-1222; 3201 New Mexico Avenue, Suite 310, Washington, DC 20016, general information: 202-362-7217; www.aapcc.org.

American College of Emergency Physicians, 1125 Executive Circle, Irving, TX 75038-2522; 800-798-1822; www.acep.org.

American Red Cross, 430 17th Street NW, Washington, DC 20006; 877-272-7337; www.redcross.org/pubs/.

Virtual Naval Hospital, U.S. Navy Bureau of Medicine and Surgery and the University of Iowa, *First Aid Manual for Soldiers,* U.S. Department of the Army manual; www.vnh.org/FirstAidForSoldiers/fm2111.html.

Food Contamination and Safety

Food and Drug Administration, Center for Food Safety and Applied Nutrition, Emergency Hotline: 301-443-1240; 200 C Street SW, Washington, DC 20204, general information: 800-532-4440; www.fda.gov.

FoodSafety.gov (U.S. government gateway to food safety information); www.food safety.gov.

U.S. Department of Agriculture, Food and Drug Administration, Foodborne Illness Education Information Center; Meat and Poultry Hotline: 800-535-4555; www .fsis.usda.gov.

U.S. Department of Agriculture, Food Safety and Inspection Service, Washington, DC 20250; www.fsis.usda.gov.

Identity Protection Services

Credit Manager (Credit Expert), P.O. Box 310, Blue Ridge Summit, PA 17214; 800-787-6864; www.creditexpert.com.

Equifax Credit Watch, P.O. Box 740241, Atlanta, GA 30374; 800-685-1111; www.equifax.com.

Identity Guard, P.O. Box 222455, Chantilly, VA 20153-2455; 800-214-4791; www.identityguard.com.

Identity Theft

Federal Trade Commission, Identity Theft Clearinghouse, 600 Pennsylvania Avenue NW, Washington, DC 20580; 877-438-4338 or 877-FTC-HELP; www.consumer.gov/idtheft/victim.htm.

Identity Theft Resource Center, P.O. Box 26833, San Diego, CA 92196; 858-693-7935; www.idtheftcenter.org.

National Association of Attorneys General, 750 First Street NE, Suite 1100, Washington, DC 20002; 202-326-6000; fax 202-408-7014; www.naag.org.

Project Money $mart, Federal Reserve Bank of Chicago, 230 S. LaSalle Street, Chicago, IL 60604-1413; 312-322-5322; www.chicagofed.org.

Social Security Administration, SSA Fraud Hotline, P.O. Box 17768, Baltimore, MD 21235; 800-269-0271 or 800-772-1213; www.ssa.gov.

U.S. Government Identity Theft Website, Federal Trade Commission; www.consumer.gov/idtheft/index.html.

U.S. Postal Inspection Service; 800-ASK-USPS; www.usps.com/postalinspectors/welcome.htm.

Information Technology Training

Computer Security Institute; 415-947-6320; www.gocsi.com.

Information Systems Security Association, Inc.; 800-370-4772 or 414-768-8000; www.issa-intl.org.

International Information Systems Security Certification Consortium, Inc.; www.isc2.org.

MIS Training Institute; 508-879-7999; www.misti.com.

SANS Institute; 866-570-9927 or 540-372-7066; www.sans.org.

TruSecure, International Computer Security Association; 888-627-2281 or 703-480-8200; www.icsa.net.

Insurance—Identity Theft

Travelers, New York; 888-695-4635; www.travelerspc.com/personal/equote/theft/.

Insurance—Kidnap/Ransom/Extortion

Aon (Canada), Toronto; 416-868-5500; www.aon.ca.

Chubb Group of Insurance Companies, Warren, N.J.; 908-903-2000; www
.chubb.com/businesses/dfi/kidnap.html.

Kemper Insurance Companies, Kemper Financial Insurance Solutions, New York,
N.Y.; toll-free 877-KEMPER-6 or 646-710-7000; www.fis.kemperinsurance.
com/kidnap.asp.

Insurance—Travel

Access America, Inc., Richmond, Va.; 800-284-8300; www.accessamerica.com.

ASA, Inc., International Health Insurance, Phoenix, Ariz.; 888-ASA-8288; www
.asaincor.com.

AXA Assistance, Bethesda, Md.; 301-214-8200; www.axa-assistance-usa.com.

Clements International, Washington, D.C.; 800-872-0067 or 202-872-0060; www
.clements.com.

Global Alert!, Van Nuys, Calif.; 800-423-3632; www.globalalerttravel.com.

Health Care Global, Wallach & Co., Middleburg, Va.; 800-237-6615, 540-687-
3166, or 540-281-9500; www.wallach.com.

Highway To Health, Fairfax, Va.; 888-243-2358; http://highwaytohealth.com.

International Medical Group, Indianapolis, Ind.; 800-628-4664 or 317-655-4500;
www.imglobal.com.

MEDEX International, Timonium, Md.; 800-732-5309; www.medexassist.com.

MultiNational Underwriters, Inc., Indianapolis, Ind.; 800-605-2282; www.mnui
.com.

Petersen International Underwriters, Inc., Valencia, Calif.; 800-345-8816; www
.piu.org.

Travelex, Omaha, Neb.; 800-228-9792; www.travelex-insurance.com.

Travel Guard, Noel Group, Stevens Point, Wis.; 800-826-1300; www.noelgroup
.com.

Travel Insurance Services, Walnut Creek, Calif.; 800-937-1387 or 925-932-1387;
www.travelinsure.com.

Unicard Travel Association, Overland Park, Kans.; 800-501-0352; www.unicard
.com.

Internet/Telemarketing Fraud

Equifax Fraud Alert, P.O. Box 740241, Atlanta, GA 30374; 888-766-0008; www
.equifax.com.
Federal Trade Commission, 600 Pennsylvania Avenue NW, Washington, DC 20580;
877-FTC-HELP; www.ftc.gov.
Internet Fraud Complaint Center (a partnership between the Federal Bureau of Investigation and the National White Collar Crime Center); www1.ifccfbi.gov.
National Fraud Information Center, P.O. Box 65868, Washington, DC 20035; 800-
876-7060; www.fraud.org.

Junk Mail—Opt Out

Direct Marketing Association, Mail Preference Service, Box 643, Carmel, NY 10512;
www.the-dma.org.

Mail Fraud and Theft

U.S. Postal Inspection Service; 800-ASK-USPS; www.usps.com/postalinspectors/
welcome.htm.

Mail Handling

Centers for Disease Control and Prevention, Emergency Response Hotline (24
hours): 770-488-7100; 1600 Clifton Road, Atlanta, GA 30333, general information: 800-311-3435, 888-246-2675, or 404-639-0385; www.cdc.gov.
Public Health Foundation, 1220 L Street NW, Suite 350, Washington, DC 20005;
202-898-5600; www.phf.org.
U.S. Postal Service, Mail Security; 800-ASK-USPS; www.usps.com.

News

U.S.-BASED NEWS SERVICES AND NEWSPAPERS

ABC News; www.abcnews.go.com.
Associated Press; http://wire.ap.org.
CBS News; www.cbsnews.com.
CNN; www.cnn.com.
Fox News; www.foxnews.com.
Los Angeles Times; www.latimes.com.

NBC News; www.msnbc.com.
New York Times; www.nytimes.com.
Voice of America; www.voa.gov.
Wall Street Journal; www.wsj.com.
Washington Post; www.washingtonpost.com.

FOREIGN-BASED NEWS SERVICES AND NEWSPAPERS

Agence France-Presse; www.afp.com.
British Broadcasting Corporation; www.bbc.co.uk.
Financial Times; www.ft.com.
Reuters; www.reuters.com.
Russia's Pravda; http://english.pravda.ru/.

DIRECTORIES OF ONLINE NEWSPAPERS

Kidon Media Link; www.kidon.com/media-link/usa.shtml.
Newslink; http://newslink.org.

DIRECTORIES OF RADIO AND TV STATIONS

Kidon Media Link; www.kidon.com/media-link/usa.shtml.
NewsDirectory.com; www.newsdirectory.com.
Newslink; http://newslink.org.
Radio-Locator; www.radio-locator.com.

ONLINE TRANSLATOR OF FOREIGN-LANGUAGE NEWS AND WEB SITES

Alta Vista's Babelfish; http://world.altavista.com.

Nuclear Hazards/Accidents/Weapons and Radiation

American Council on Science & Health, 1995 Broadway, Second Floor, New York, NY 10023-5860; 212-362-7044; www.acsh.org.
Center for Nonproliferation Studies, Monterey Institute of International Studies, 460 Pierce Street, Monterey, CA 93940; 831-647-4154; http://cns.miis.edu.
Federal Emergency Management Agency, 500 C Street SW, Washington, DC 20472; 800-462-9029; www.fema.gov/hazards/nuclear/.
Federation of American Scientists, 1717 K Street NW, Suite 209, Washington, DC 20036; 202-546-3300; www.fas.org.

Radiation Emergency Assistance Center, Oak Ridge Institute for Science and Education, P.O. Box 117, MS 39, Oak Ridge, TN 37831-0117; 865-576-3131; www.orau.gov/reacts/intro.htm.

U.S. Nuclear Regulatory Commission (including potassium iodide information), Washington, DC 20555; 800-368-5642 or 301-415-8200; www.nrc.gov/what-we-do/radiation.html.

Report Terrorist/Suspicious/Criminal Activity

Federal Bureau of Investigation, J. Edgar Hoover Building, 935 Pennsylvania Avenue NW, Washington, DC 20535-0001; 202-324-3000; www.fbi.gov (local FBI office locator at www.fbi.gov/contact/fo/fo.htm).

National Infrastructure Protection Center, NIPC Watch and Warning Unit, J. Edgar Hoover Building, 935 Pennsylvania Avenue NW, Washington, DC 20535-0001; 202-323-3205 or 888-585-9078; nipc.watch@fbi.gov or www.nipc.gov/incident/cirr.htm.

Royal Canadian Mounted Police (phone number and department locator); www.rcmp-grc.gc.ca.

U.S. Nuclear Regulatory Commission, Nuclear Incident Hotline: 301-816-5100; Washington, DC 20555; 800-368-5642 or 301-415-8200; www.nrc.gov.

Sexual Assault Prevention

National Crime Prevention Council, 1000 Connecticut Avenue NW, 13th Floor, Washington, DC 20036; 202-466-6272; www.ncpc.org.

U.S. State Department, Bureau of Diplomatic Security; http://ds.state.gov.

Smoke Hoods

Air Security International; 713-430-7300; www.airsecurity.com.

Brookdale International Systems; 800-459-3822 or 604-324-3822; www.evac-u8.com or www.smokehood.com.

Social Security Fraud

Social Security Administration, SSA Fraud Hotline, P.O. Box 17768, Baltimore, MD 21235; 800-269-0271 or 800-772-1213; www.ssa.gov.

Telemarketing—Opt Out

Direct Marketing Association, Telephone Preference Service, Box 643, Carmel, NY 10512; www.the-dma.org.

Terrorism Background Information

Brookings Institution, 1775 Massachusetts Avenue NW, Washington, DC 20036; 202-797-6000; www.brook.edu.

Cato Institute, 1000 Massachusetts Avenue NW, Washington, DC 20001-5403; 202-842-0200; www.cato.org.

Center for Strategic and International Studies, Homeland Defense, 1800 K Street NW, Suite 400, Washington, DC 20006; 202-887-0200; www.csis.org/burke/hd/index.htm.

Centre for Defence and International Security Studies, Department of Politics and International Relations, Cartmel College, Lancaster University, Lancaster LA1 4YL, United Kingdom; 44-0-1524-594-261; www.cdiss.org/terror.htm.

Council on Foreign Relations, Harold Pratt House, 58 E. 68th Street, New York, NY 10021; 212-434-9400; www.cfr.org.

Dudley Knox Library, Naval Postgraduate School, 411 Dyer Road, Monterey, CA 93943; http://web.nps.navy.mil/~library/terrorism.htm.

Heritage Foundation, 214 Massachusetts Avenue NE, Washington, DC 20002-4999; 202-546-4400; www.heritage.org.

Hoover Institution, Stanford University, Stanford, CA 94305-6010; 877-466-8374 or 650-723-1754; www-hoover.stanford.edu.

International Policy Institute for Counter-Terrorism, Interdisciplinary Center Herzlia, P.O. Box 167, Herzlia 46150, Israel; fax 972-9-9513073; www.ict .org .il/.

RAND, 1700 Main Street, P.O. Box 2138, Santa Monica, CA 90407-2138; 310-393-0411; www.rand.org.

U.S. Central Intelligence Agency, Washington, D.C. 20505; 703-482-0623; www.odci.gov/terrorism/.

U.S. Defense Department, The Pentagon, Washington, D.C. 20301; www.dod.gov.

U.S. State Department, 2201 C Street NW, Washington, DC 20520; 202-647-4000; www.state.gov.

Terrorism Warnings (Online)

Australian Department of Foreign Affairs and Trade; www.dfat.gov.au.
Australian Embassy, Washington, D.C.; www.austemb.org.
Australian Federal Government; www.fed.gov.au.

Australian Prime Minister's Office; www.pm.gov.au.

British Foreign & Commonwealth Office; www.fco.gov.uk.

British Home Office; www.homeoffice.gov.uk.

British Prime Minister's Office; www.pm.gov.uk.

Canadian Government Homepage; http://canada.gc.ca.

Centers for Disease Control and Prevention; www.cdc.gov.

Royal Canadian Mounted Police; www.rcmp-grc.gc.ca.

Scotland Yard (London Metropolitan Police); www.met.police.uk.

U.K. Resilience (U.K. government antiterrorism clearinghouse); www.ukresilience .info.

U.S. Central Intelligence Agency; www.cia.gov.

U.S. Defense Department; www.defenselink.mil or www.dod.gov.

U.S. Federal Bureau of Investigation; www.fbi.gov.

U.S. FirstGov (U.S. government information gateway); www.firstgov.gov.

U.S. Office of Homeland Security; www.whitehouse.gov/homeland/.

U.S. State Department; www.state.gov.

Travelers' Health

American Society of Tropical Medicine and Hygiene, 60 Revere Drive, Suite 500, Northbrook, IL 60062; 847-480-9592; www.astmh.org.

Centers for Disease Control and Prevention, Travelers' Health Information, 1600 Clifton Road, Atlanta, GA 30333; 877-FY1-TRIP; www.cdc.gov/travel.

International Association for Medical Assistance to Travellers, 417 Center Street, Lewiston, NY 14092; 716-754-4883; or 40 Regal Road, Guelph, Ontario, Canada N1K 1B5; 519-836-0102; www.iamat.org.

International Society of Travel Medicine, P.O. Box 871089, Stone Mountain, GA 30087-0028; 770-736-7060; www.istm.org.

Public Health Resources: State Health Departments, Centers for Disease Control and Prevention; www.cdc.gov/mmwr/international/relres.html.

Travel Health Online; www.tripprep.com.

Travel Emergencies

Canadian Department of Foreign Affairs; Canadian citizens overseas can call collect 613-996-8885; for calls originating in Canada or the United States, use 800-267-6788 or 613-944-6788; www.voyage.gc.ca.

International Association for Medical Assistance to Travellers, 417 Center Street, Lewiston, NY 14092; 716-754-4883; or 40 Regal Road, Guelph, Ontario, Canada N1K 1B5; 519-836-0102; www.iamat.org.

U.S. Department of State, U.S. Citizens and Families Helplines (including cases of arrest abroad); 888-407-4747, 202-647-5225, 317-472-2328, or 202-647-4000; www.state.gov.

Travel Warnings

British Foreign & Commonwealth Office, Foreign Travel Advisories; 020-7008-0232/0233; www.fco.gov.uk/travel.

Canadian Department of Foreign Affairs, Foreign Travel Advisories; www.voyage.gc.ca or www.voyage.gc.ca/destinations/menu_e.htm.

Federal Aviation Administration; 800-221-0673; www.faa.gov.

U.S. Department of State, Travel Hotline: 202-647-5225; www.state.gov.

U.S. Border Crossings Wait Times

U.S. Customs Service, Travel Information; www.customs.gov.

U.S. Customs Regulations

U.S. Customs Service, Traveler Information; www.customs.gov.

U.S. Embassies and Consulates

Directories of U.S. embassies and consulates worldwide: http://travel.state.gov/links .html or http://usembassy.state.gov.

U.S. Passport Information

U.S. State Department, National Passport Information Center; telephone billing fees are charged for calls to 900-225-5674; Visa, MasterCard, and American Express credit holders can charge calls to 888-362-8668; http://travel.state.gov/passport _services.html.

Victims of Terrorism/Crime

Office for Victims of Crime, U.S. Department of Justice; 800-627-6872; www.ojp .usdoj.gov/ovc/.

Terrorism Victim Hotline, U.S. Department of Justice; 800-331-0075.

Visa Services

U.S. State Department; 202-663-1225; http://travel.state.gov/visa_services.html.

Water Security

U.S. Environmental Protection Agency, Environmental Emergency Hotline: 800-424-8802; 1200 Pennsylvania Avenue NW, Washington, DC 20460; 215-814-5000; www.epa.gov.

INDEX

ABOUT THE AUTHOR

Juval Aviv has an MA in business from Tel Aviv University and served as an officer in the Israel Defense Force (major, retired) leading an elite commando/intelligence unit. Additionally, Mr. Aviv was selected by the Israel Secret Service (Mossad) to participate in a number of intelligence and special operations in many countries in the late 1960s and 1970s. In 1984, a true account of one mission was published; the book became a best seller and was later made into an award-winning film.

Over the past 20 years, Interfor, Inc., founded by Mr. Aviv, has become a leader in corporate intelligence worldwide, working with U.S. and foreign law firms, major banks, insurers, and governmental agencies. Interfor's areas of expertise include security and vulnerability assessments, industrial espionage inquiries, due diligence investigations, litigation support, and competitor intelligence. Interfor also provides asset-searching and corporate-fraud investigation services, both in the United States and internationally, in cases involving large debts or where assets have been hidden offshore in bank-secrecy jurisdictions to avoid judgment creditors.

For over 30 years Mr. Aviv has worked with corporations and other entities, both domestically and internationally, on security measures for the protection of assets and personnel. Mr. Aviv's experience is broad and includes high-profile clients such as El Al Airlines. While working as a consultant with El Al, Mr. Aviv surveyed the existing security measures in place and updated and developed El Al's security program, making El Al the safest airline in business today.

Mr. Aviv is a leading authority on terrorist networks and their inner workings and served as lead investigator for Pan Am Airways into the Pan Am 103–Lockerbie terrorist bombing. Mr. Aviv has been a speaker and panel participant for many organizations, including the International Bar Association, the National Conference of Bankruptcy Judges, the U.S. Department of Justice, the Internal Revenue Service, and the American Bankruptcy Institute.